HOW TO FORGE A FROGMAN

A RECRUIT'S ACCOUNT OF BASIC TRAINING IN SINGAPORE'S NAVAL DIVING UNIT

MAX WEST

Marshall Cavendish
Editions

© 2017 Maximillian Alisdair West

Cover design by Aneirin Flynn
Photographs by the author
Book design by Benson Tan

Published by Marshall Cavendish Editions
An imprint of Marshall Cavendish International

Other Marshall Cavendish Offices:

Marshall Cavendish Corporation. 99 White Plains Road, Tarrytown NY 10591-
9001, USA • Marshall Cavendish International (Thailand) Co Ltd. 253 Asoke, 12th
Flr, Sukhumvit 21 Road, Klongtoey Nua, Wattana, Bangkok 10110, Thailand •
Marshall Cavendish (Malaysia) Sdn Bhd, Times Subang, Lot 46, Subang Hi-Tech
Industrial Park, Batu Tiga, 40000 Shah Alam, Selangor Darul Ehsan, Malaysia

Marshall Cavendish is a registered trademark of Times Publishing Limited

National Library Board, Singapore Cataloguing-in-Publication Data

Name(s): West, Max.
Title: How to forge a frogman : a recruit's account of basic training in Singapore's
Naval Diving Unit / Max West.
Other title (s): Recruit's account of basic training in Singapore's Naval Diving Unit
Description: Singapore : Marshall Cavendish Editions, [2017]
Identifier(s): OCN 983214340 | ISBN 978-981-4721-75-2 (paperback)
Subject(s): LCSH: Basic training (Military education)--Singapore. | Underwater
demolition teams--Singapore. | Singapore. Navy--Officers.
Classification: DDC 359.965095957--dc23

Printed in Singapore by Markono Print Media Pte Ltd

CONTENTS

Disclaimer

....

This publication does not represent the views of the Singapore Armed Forces, the Ministry of Defence, or the Republic of Singapore or any of its governing bodies. The opinions expressed are those of the author alone.

The names of all regular military personnel and nearly all National Servicemen have been changed for their protection and privacy. Their pseudonyms are entirely fictitious and bear no reflection on actual persons.

And, as we all know, people in the Navy do not swear.

AUTHOR'S NOTE

By law, upon completion of their secondary education, all Singaporean men are required to perform 22 months of National Service (NS). At the start of NS, every combat-fit recruit undergoes nine weeks of Basic Military Training (BMT) before being posted to various vocations across the Singapore Armed Forces.

The following account of my BMT was handwritten at the time of its occurrence, day by day, painstakingly and often furtively. Only later was it typed and transcribed for publication. No material was recreated in hindsight.

This is one recruit's story.

A (BRIEF) GLOSSARY

....

Ang moh (n.): Literally, "red head." Generally refers to anyone who appears Caucasian

Chao keng (v.): To feign injury for the purposes of evading training or responsibilities

mee soto (n.): A noodle soup dish common in Singapore, Malaysia and Indonesia.

Milo (n.): A chocolate malt beverage popular in Singapore.

Pang kang (v.): To be done with work; to be free to go

Pasar malam (n.): Literally, "night market." A street market that opens in the evening, typically around residential neighborhoods

DRILL COMMANDS

Sedia: At attention

Senang diri: At ease

Kekanan pusing: Right turn

Kekiri pusing: Left turn

Diam: Strictly at attention; no movement

Dari kiri: By the left

Cepat jalan: Quick march

Hentak kaki: Quick mark time (march in place)

Julang senjata: High port arms

Berhenti: Halt

The author with his parents on enlistment day, March 11, 2013.

WEEK
ONE

MONDAY, MARCH 11, 2013

Today, on the 11th of the month of March, I enlisted into National Service. I joined the Naval Diving Unit. We're in Sembawang Camp.

There are 100 of us.

I'm nervous.

IF I MAY

My name is Maximillian, but I go by Max. My father is American and my mom is Singaporean Chinese. I was born and raised in Singapore. I'm 180 cm tall, and weigh 78 kg.

I like to write.

NAVAL DIVING WHAT?

The Naval Diving Unit is an elite formation of the Singapore Armed Forces. In peacetime, NDU conducts maritime security and counter-piracy operations, underwater search and salvage missions, humanitarian assistance, and disaster relief.

NDU only has two intakes of National Servicemen each year, each consisting of fewer than 120 enlistees. I wanted to come here, and was lucky to be chosen.

We are the 39th Batch.

THE FIRST TEST

We had a swimming lesson on Wednesday. We had to wear SAF standard-issue swimwear, which are navy blue triangle trunks. I've never seen so many wedgies in my life. On the bright side, I passed the Category 1 swimming test, our first official evaluation. It's a test of water confidence.

Ivan Tan, a national runner, and I were the first to attempt the test. We donned our number fours, green camouflage-patterned uniforms. The test consisted of a 50-m breaststroke swim, followed by five minutes of treading water. It's untimed, so it makes sense to swim slowly and conserve energy. I was nervous, though, and rushed through the breaststroke, wasting energy. When I started treading water, I was already out of breath. The five minutes felt longer than they should have, but I made it through.

Next, I removed my number four and used the pants to make a float. I tied the legs of the pants into knots and flipped them over my head, trapping air inside. I struggled a bit with the pants inflation, but in the end, I managed to pass. So did Ivan.

The Category 1 test has been our only evaluation so far.

THE SAF'S NEXT TOP MODEL

Looking around during the Cat 1 test, I noticed a strange assortment of physiques. NDU is one of the fittest units in the SAF, so I thought everybody would be in shape.

Surprisingly, some guys are chubby, and some are really skinny. A few can't even do one proper push-up or pull-up. On the other hand, many are ripped. A handful are bodybuilders, but most are lean and toned.

Nobody wears glasses, which is highly unusual for one hundred Singaporeans. Perfect eyesight is a requirement for naval divers.

⋅→▰◉ ⎯⎯ ◉▰←⋅

There are only two members of 39 who aren't fully Asian. As I'm half-American, I'm one of them. The other is Aneirin Flynn, who's half-British. On enlistment day, I saw him and said, "Hey, you're white, I'm white. Wanna be friends?"

Before enlisting, I thought being half-American would hurt me. I thought I'd be picked on to do extra work or tougher training, and that it might be tougher to make friends. In truth, it's been the opposite. The novelty of being Eurasian has helped me get to know people. Nearly everyone wants to know what an *ang moh* is doing in NS. I'm also the only one who didn't study at a local school. I attended Singapore American School, an international school under the American curriculum.

Aneirin and I stand out. Unexpectedly, this has been an advantage. The instructors have noticed us, and most of them like us.

BUT WE HARDLY KNEW YOU

After just seven days, Batch 39 has already lost two men. One went out-of-course (OOC) because of sinusitis and flat feet. The other went OOC because he tripped while walking into the cookhouse and fractured his arm.

Our class's total strength has fallen to 98.

WEEK
TWO

SATURDAY, MARCH 16 - SUNDAY, MARCH 17

Typically, we'll book out on Friday nights and book in on Sunday nights. However, the first two weeks of NS are the "confinement period," so we stayed in camp over the weekend.

Let the second week begin.

MONDAY, MARCH 18

This week, Aneirin and I have volunteered to be ICs. I won't lie — I did this purely to increase my chances of going to Officer Cadet School.

There are always two class ICs. IC stands for In-Charge. These two poor dudes are tasked with running the class. The ICs rotate weekly.

This morning, our Platoon Commander, 2nd Warrant Officer (2WO) Foo, asked for two volunteers to be this week's ICs.

Last night, Aneirin and I had agreed to volunteer to be ICs together. Now, as Warrant Foo's question hung over us, I looked at Aneirin and froze, paralyzed by nervousness. I shook my head. Aneirin smiled, looked away, and raised his hand.

Shit, I thought. I raised my hand too.

Warrant Foo was surprised. "You two want to be ICs? Good," he said.

So, here we are.

The ICs are responsible for knowing the class head count and keeping the batch on schedule. As I'm starting to find out, this is harder than it sounds. Getting nearly a hundred dudes to behave and cooperate is a task worthy of a Greek tragedy. Even worse, the ICs get punished for the class's mistakes. For instance, whenever the batch is late, Aneirin and I hold the push-up position until everyone arrives.

The ICs also give the drill commands. The commands are in Malay and must be delivered in a precise order. I've really been struggling with these. My pronunciation is terrible, and I can't get the accent right. A few times, instead of following my commands, the batch has just laughed at me. I'm taking each screw-up deeply. I want to succeed, but each time I say the wrong thing in front of the class, I tense up.

Aneirin is able to laugh off his mistakes. When he mispronounces something or forgets the command, he laughs and yells, "Shit!" Everyone laughs with him. When I mess up, I become serious, and everyone notices. I need to unearth the class's respect.

We had a stand-by bed this afternoon, conducted by Master Sergeant (MSG) Dennis, one of our Platoon Sergeants. Stand-by beds are area inspections. During the inspection, the batch stands at attention outside their cabins, and the instructors inspect the rooms. The class ICs follow the instructors around.

So far, stand-by beds are the worst part of military life. I can't stand them. In addition to the room being spotless, all of our belongings must be arranged in a precise way. We place our pillows over our blankets, have our towels folded at the edge of the bed, fold and place our number fours on the cabinet's top left shelf, fold and place our PT shirts over our PT shorts next to our socks on the cabinet's bottom left shelf, place toothbrushes next to toothpaste next to razors next to cups on the cabinet's top right shelf, and lay our field packs and helmets on top of the cabinet. Everything else we own is jammed into our duffel bags, which are then stacked neatly in a corner. The cabinet arrangement leaves about 90% of the available space empty. As a result, our duffel bags are crammed 100% full of our shit.

Our standard in the stand-by bed was subpar. As ICs, Aneirin and I were held responsible. Master Dennis punished us, putting us through set after set of push-ups and jumping jacks while the rest of the batch watched.

I earned some much-needed respect for that. I felt it evaporate that night, though, when Aneirin told me, "Everyone says you should chill out a bit."

I don't think I've been overly hard on the class. Yes, I yell when they're unresponsive, but only after asking nicely hasn't worked. I haven't figured out how to simultaneously be nice to the class and get them to listen. When we're unsupervised, everyone just wants to mess around, and it takes forever to get their attention. Perhaps I should be more lenient. But I'm just doing my best to keep us on time and prevent us from getting punished.

I almost always feel unsure of myself. My old insecurities are creeping back in.

THE PERKS

Being IC can be rewarding, though. There's the pre-eating chant, for instance.

During meals, we always wait for everyone to collect their food and then start eating together. Every man stands behind his chair, food untouched, until everyone but the ICs have gotten their plates. This usually takes about 15 minutes. By now, having stared at the food in front of them for a quarter of an hour, everyone's twitching with hunger.

When the ICs finally have platefuls, they stand before the class and shout, "Ready?" The class roars back, "READY!" The ICs continue, "Our core values!" The class hollers the Naval Diving Unit's three core values: "HONOR, INTEGRITY, TEAM SPIRIT!"

The ICs yell, "Eat up!" The class echoes the call and everyone can eat at last. This is always the most full-bodied cheer of the day.

Overall, I'm getting along well with the class, though being IC has altered some friendships. It's a different relationship dynamic, as now they have to listen to me. I also believe my struggles, especially with the drill commands, have caused some to doubt my leadership. I guess I'll have to change their minds.

NAVY TALK

The Navy has a unique vernacular. When we say a number, we say its individual digits. So, when people say "Batch 39," they don't say "thirty-nine." They say "three-nine."

Our bunks are "cabins."

Toilets are "heads."

To fall in is to "muster." We do this in rows of three.

Physical activities and exercises are "evolutions."

Naval divers are known as "frogmen." Just inside the NDU gate is a statue captioned "Frogman from the Sea." The frogman holds a trident in one hand and a bomb in the other. It has one foot in the sea and one foot on land. Right now, we aren't frogmen. We're tadpoles.

CURRENT AFFAIRS

One week has passed. Life in NDU hasn't been what I expected. NDU has a reputation for intense training; I pictured muscle-crushing workouts and mind-melting runs. Instead, in the first week, we've only had one physical training (PT) session and two swimming lessons. There's been lots of administrative work and in-processing. We received our uniforms and equipment. We were assigned bunks. We've endured talks from an endless flow of people, from instructors to insurance salesmen.

I'm still waiting for the pain to begin.

TUESDAY, MARCH 19

This morning we took the IPPT, the Individual Physical Proficiency Test. The IPPT standards for the Naval Diving Unit are far tougher than the typical SAF standard. To pass, we need Diver's Gold, which is way harder to get than regular IPPT gold. To score Diver's Gold, you have to score five points in every category, and run 2.4 km in 9 minutes and 14 seconds or less. That means we must meet or exceed the following:

- 12 pull-ups,
- 43 sit-ups,
- 243 cm standing broad jump (SBJ),
- 10.1-second shuttle run,
- 9:14 minutes or faster for the 2.4-km run.

I was confident in each station except for the 2.4-km run. I scored gold on the NAPFA, the pre-enlistment physical test, running 9:17. That was a long time ago, though, and I'm afraid my fitness has declined since then.

Yesterday, the class was divided into two platoons, aptly named Platoon 1 and Platoon 2. Each platoon has four sections. I'm in Platoon 1 Section 4.

The IPPT began. We split into sections and cycled

through the stations. I did exactly 12 pull-ups and 43 sit-ups, stopping at the minimum Diver's Gold standard to conserve energy. I ran the shuttle run in 9.4 seconds.

Aneirin is pretty fit. His personal best in the 2.4 km is around 9:10. Surprisingly, though, he's can't jump. As I waited in line at the pull-up station, I heard someone boom, "West, get over here!"

1st Warrant Officer (1WO) Jaya, our Training Officer, summoned me to the standing broad jump station. I hurried over. Aneirin was preparing to jump.

"Flynn, tell West what you just told me," Warrant Jaya said.

Aneirin laughed. "I said, sir, that white men can't jump," he said jokingly.

I laughed. "What, Flynn, you can't make 243?"

Warrant Jaya said, "No, he cannot, and he says the reason why is because white men can't jump. Go ahead, West. Let's see if you can."

The SBJ is my strongest station. I grinned and jumped 265 cm. Warrant Jaya's eyes widened. He said, "Flynn, you still want to tell me white men can't jump?"

"No, sir."

Warrant Jaya waved to Warrant Foo and Master Dennis. "Hey, come here," he called. "West, jump one more time."

By now, a crowd of batch boys had gathered to watch. I smiled and jumped 275 cm. The class ooh-ed and ahh-ed.

That jump was important for me. The instructors

and a lot of the batch saw it. Hopefully it'll help me stand out.

The 2.4-km run was last. Unfortunately, four days before enlisting, I hurt myself. I was jogging in the Botanic Gardens when I felt sharp pains in my right hip. A physiotherapist told me I have irritation in the joint in my right gluteus muscle. Since then, I have occasional sharp pains while walking, and I haven't attempted running.

I decided that if my hip hurts during the run, I would drop out. I don't want to risk a serious injury in the second week. I guessed there was at least a 50% chance I wouldn't be able to finish the run.

The Naval Diving Unit doesn't have a standard running track. Instead of running six laps of a 400-m track, we run four laps of a 600-m route around our camp.

So far, I was on track to score Diver's Gold. All I had to do was run 9:14 or faster.

Nearly 50 of us toed the start line. I edged my way to the front. My pulse thumped in my eardrums, my heart flying.

An instructor with a stopwatch called out, "3, 2, 1, go!"

I planned on trying to maintain a consistent pace throughout the run, but at the start, nearly everyone immediately began sprinting. I started in the middle of the pack, but gained steadily as other runners slowed down. By the end of the second lap, the halfway point, Aneirin and I were in the lead, running alongside each other. I kept waiting for a shooting pain to slice through my right hip, but it never arrived. Miraculously, I was pain-free.

Aneirin and I pushed on side-by-side, tied for the lead, until the start of the final lap. As we began the final round, Lieutenant (LTA) Jackson sprinted ahead, shouting, "Come on, come on!" Aneirin took off, running alongside LTA Jackson. I was nearly burned out. I worried that if I sprinted now, I would run out of steam before the finish line. I continued at my pace.

"Come on, Max, catch up!" Jackson called from ahead, but I couldn't.

I ran hard the rest of the way and finished second in the platoon with a time of 8:50. The only guy who beat me? Aneirin, who ran 8:41.

Aneirin wasn't able to notch Diver's Gold, however. He fell short in pull-ups and in the standing broad jump. But that white boy can run.

In Platoon 1, only Raymond Hoe and I got Diver's Gold. In Platoon 2, four more joined us—Ivan Tan, Elston Sam, Shawn Ong, and Peter Wong. Ivan used to be a national runner. Unsurprisingly, he's fast as hell. He ran 8:03 today. His personal best is around 7:40.

Ninety-four of us took the IPPT today. Six reached Diver's Gold.

The class is gaining a bit of respect for me. I've eased

up a little, and Aneirin has realized that the class won't listen unless we yell. Now, he's raising his voice more often so I don't have to do all the shouting myself.

◦▩ ⎯⎯ ▣◦

In the afternoon, we were addressed by Deputy Commander NDU, a Senior Lieutenant Colonel (SLTC). He's second in command after Commander NDU, a full Colonel (COL).

Dy Commander asked if there were any national athletes in the group. We all assumed he was curious if there were any great runners or swimmers.

A guy named Zhi Heng stood up.

Dy Commander nodded. "What sport do you play?"

Zhi Heng responded, "Bowling, sir."

RESPONSIBILITY ISN'T FOR EVERYONE

On the first day, we were all issued Swiss Army knives. I've already lost mine. Shit.

WEDNESDAY, MARCH 20

We did a recovery run this morning, jogging 4 km in 25 minutes. We followed that with some static bodyweight exercises, focusing on push-ups and core work. In the afternoon, we went for a swim. We've begun our training for drown-proofing.

Drown-proofing is a series of water proficiency and confidence tests carried out with our hands and feet tied behind our backs. It's a veto factor, meaning those who do not complete it will go out-of-course.

FOCUS

Throughout these first weeks, I've been scatter-brained. I constantly forget where I've put things down and waste time trying to find them. I think it's because I'm getting less sleep than I'm used to. We wake up at 0530 or earlier every morning.

The ICs carry around a briefcase holding the class's 11Bs, military identity cards. Today, I left the briefcase unattended by the pull-up bars. Losing your own 11B is a severe offense. I can't imagine the punishment I would've gotten for losing the entire class's 11Bs. Fortunately, Aneirin found the briefcase.

Then, tonight, I forgot to remind 39 to muster

because I was taking a shit. As a result, we were late for dinner. I honestly don't know what's wrong with me. I'm not usually like this.

During dinner, Aneirin told me, "Hey man, you need to get your shit together." He later apologized, but I thanked him. I needed to hear it.

COMMUNICATION BREAKDOWN

Aneirin and I are often misunderstood because of our American and British accents. Since I've become IC, there have also been jokes about whether this is the Singapore Navy or the US Navy SEALs.

Aneirin was the only white guy at his local school, so he's used to this. I'm not quite there yet.

UPON FURTHER OBSERVATION

Now that I know it better, here's a closer look at the Naval Diving Unit.

NDU is located inside Sembawang Camp. The main NDU compound houses buildings, offices, equipment storage, the pool, the football field, and the cookhouse. Our pool is currently under repair, so when we have swimming evolutions, we take a 20-minute bus ride to Nee Soon Camp or Khatib Camp. The bus rides serve as naptime. They're heavenly.

Each building in NDU is numbered. Our dormitory is Block 29, a rectangular chunk of white concrete. It was built by the British when they occupied Singapore.

In accordance with Navy vernacular, we refer to 29 as "two-nine." 29 is about a half-kilometer away from the main compound.

Though Block 29 is actually a four-story building, the first floor is called the ground floor, the second floor is called the first floor, and so on. There are no cabins on the ground floor. Levels 1–3 have 10 cabins each. I stay in cabin #02-03, with my three cabinmates.

THE CABINMATES...

Benjamin is a good guy who laughs at my jokes. I like him. He was a competitive rower at Hwa Chong and is deceptively strong. He weighs less than 70 kg, but bench presses 90. He has been unable to participate in physical activity so far because of a strained forearm, which he injured in an MMA [mixed martial arts] class prior to enlisting.

Noel is a slightly awkward jokester. He and Benjamin knew each other before enlisting. They've been pals for years. Noel is skinny and struggles with push-ups and pull-ups, but he's a competitive sprinter. He's fast as hell and his calves are monstrous.

Thomas is the quietest of the four of us. I thought he was unfriendly at first, but he's just naturally reserved. Now that I know him, it's clear he's a nice guy who can also be funny. Thomas is the only vegetarian in the batch. He's never had a mouthful of meat in his life. He's also in shape, with a 2.4-km time verging on Diver's Gold.

. . . AND THE CABIN

Our cabin is modest, with two double-decked bunk beds, four cabinets, a table, and a couple of chairs. I snagged a top bunk above Thomas. Benjamin sleeps on the other top bunk over Noel.

Each cabin has a small balcony in the back for us to dry our clothes. The balconies overlook a small field behind Block 29, which is home to three huge trees. These trees are perpetually shedding leaves, which we're perpetually picking up. Every morning, we do a half hour of "area cleaning," which consists solely of collecting leaves. This is probably the silliest portion of our training so far.

Directly opposite Block 29, separated by a small parade square, is Block 29A. This is home to Batch 38. The 38th Batch is nearing the end of their Combat Diver Course. They are approaching Hell Week. I'd like to ask them about life in NDU, but we were ordered not to approach them. They probably wouldn't want to talk to us anyway.

REFLECTIONS

Warrant Foo told us that throughout our time in NDU, we'll receive "deals"—offers that reward good performance. The main incentive is usually time off. The system has proved an excellent motivator. As Warrant Jaya said, "During my time in the SEAL course, our instructors had a term for this. They called it, 'Good deal for good SEAL.'"

NDU's training is supposedly the toughest in all of National Service. Yet, to date, it hasn't been too challenging. Our instructors and officers have stressed that our training will be progressive, getting more difficult over time.

I'm wondering if NS will be as tough as I expected. I'm not sure if I want it to be. I'm one of the fitter recruits, and as the training is the same for everyone, I don't have to push myself as hard as others do. It's tempting to hope that the training will remain modest.

But I'm also tempted to hope that NDU will show me who I am. I expected tremendous physical and mental trials, searing lashes that would strip me of my falseness and, for the first time, reveal what's at my core, raw and bared. I want to see who I am beneath the veneers. I tell myself I'm a fighter, a winner, a warrior. But I won't know until I need to find out.

WEEK
THREE

SUNDAY, MARCH 21

Last Friday, we booked out for the first time, heading home for the first time since enlisting. We booked in tonight. We were told to reach our cabins by 2100. We have to book in together, so we met outside the NDU gate at 2000 to walk in.

The weekend just flew past. It's incredible to think that over one-fifth of BMT is over.

We reached our dorms at 2045. 1st Sergeant (1SG) Choi was waiting for us. He decided to check our field packs for banned items.

There's a long list of stuff we aren't allowed to bring in. The most commonly-smuggled items are phone chargers and outside food.

ONE-BAR DREAMS

Along with many others, I share the goal of going to Officer Cadet School. OCS is a nine-month course. Those who complete it are commissioned as 2nd Lieutenants.

Overall, the top 10–15% of National Servicemen are sent to OCS. However, we've been told that of the 98 members in Batch 39, only two or three of us will be selected for OCS. An additional six or so will become sergeants and the rest will be corporals. So, though naval divers endure the toughest training, we have the lowest chances of becoming officers. I don't understand why.

Right now, though, OCS is a long way away. We currently hold the rank of Recruit. As Master Dennis likes to remind us, we're "the lowest life forms in NDU."

HELLO RESPONSIBILITY, DIDN'T SEE YOU THERE

I found my Swiss Army knife. I'd stashed it in one of my boots. Oh yeah, baby.

MONDAY, MARCH 25

Aneirin and I have officially relinquished our IC duties. Our replacements, whom we appointed, are Ivan Tan, the national runner, and Zhi Heng, the proud bowler. Ivan wants to go to OCS, so he's happy to be chosen. Zhi Heng is nervous about the role.

Aneirin and I have become closer. The shared hardship of being IC brought us together. We were dropped so often that we joked the push-up position was our natural state. Whenever anyone needed anything, shouts of "IC, IC!" or "Flynn!" or "Max!" rang out. We're glad it's done, but we're also glad we did it. The instructors and officers have noticed us.

One night, when the two of us were walking somewhere, Aneirin turned to me and said, "Through thick and thin, buddy."

⊷⊗⊷ ⸻ ⊷⊗⊷

In addition to being an *ang moh*, Aneirin receives extra attention because he had a perfect score for his 'A' Levels, earning seven As. He was even interviewed by a local newspaper. In the article, he talked about enlisting in NDU, gaining publicity for the unit and impressing our officers. The batch already thinks he's ticketed for OCS.

This afternoon, Aneirin told me with a wide-eyed smile, "You know what's funny? I didn't actually get 7As."

I stopped walking. "You what?"

"I got all As, yeah. But I didn't take seven subjects. Some people take seven, but I only took six."

I said, "So, basically, you're a lying sack of shit."

He grinned. "Yeah, basically."

A few curious batch boys have asked about my school results, wanting to compare me to Aneirin. I didn't take 'A' Levels, but I took the Scholastic Aptitude Test, or SAT, an American university entrance exam. I scored 2350 of a possible 2400, putting me in the top 1%. Our instructors don't know this, however, so Aneirin still holds most of their focus. I'm only a little jealous.

MAN AND STEEL

After dinner, we had our weapons presentation, a ceremony in which every recruit receives a SAR 21 rifle, a weapon that will stay with us throughout BMT. When we reached the site of the presentation, I was struck by its serenity. Night had fallen. The parade square was dark, the shadows punctured only by torch and candlelight. Our instructors stood neatly before us. Rows of SAR 21 rifles lined the tables we'd set up earlier. The Singapore flag hung watchfully in the background.

One by one, they called our names. When mine was announced, I shouted, "Sir!" and stepped forward.

Warrant Jaya extended my rifle. Grasping a firearm for the first time in my life, I realized I was becoming a soldier. National Service isn't a formality. It's not something done for appearances. It's a serious military commitment. We really will become soldiers, and if Singapore goes to war, we really will fight.

When everyone received their rifle, we proceeded to the NDU jetty, overlooking the night sea. I realized that my rifle is far more than crafted metal and black grease. It is strength, freedom, confidence, security. Prior to this moment, I wasn't a soldier; I was a boy cast in a soldier's uniform and made to run and march and swim. The lynchpin of a soldier's identity is his weapon. Without it, he cannot fight. He cannot defend or kill. As I pulled the SAR 21 to my chest, I knew the weapons presentation was far more than the distribution of rifles to recruits. I understood why it was executed with such solemnity. For the first time, I felt like a real soldier.

The officers told us half-jokingly that our rifles will be our wives from here on out. We were encouraged to name them.

I decided to name my rifle Erica, after my first ex-girlfriend.

THE BRASS

We have a multitude of officers and instructors above us. It's taken me two weeks, but I've figured out the chain of command.

We have two Platoon Sergeants. MSG Dennis runs Platoon 1, and MSG Martin runs Platoon 2. Above them is our Company Sergeant Major (CSM), MSG Raj. Higher up is our Platoon Commander, 2WO Foo. Yet above him is our Training Officer, 1WO Jaya.

Warrant Jaya is hilarious. He'll demolish a recruit for asking a stupid question, then follow it up with a funnier anecdote. His talks always feature a peculiar hand gesture — his palms out, he'll shake his hands as if waving goodbye. He does this often enough that his nickname among the batch is "DJ Jaya." Everyone likes Warrant Jaya. He's completed the US Navy SEAL course, a fact that commands our respect.

Warrant Foo is short but strong. His stated self-opinion is that he has an enviable "charisma." It's clear that he cares for us.

Master Raj is the most gentle of our instructors. He's soft-spoken, and has yet to drop or even shout at us.

Master Dennis is the most casual of our instructors. He spends a lot of time with us. He's as funny as Warrant Jaya, and equally well-liked. It takes a lot to anger him, but once he's mad, he can be vicious.

THE HOTSHOTS

There are also two commissioned officers, LTA Jackson and 2nd Lieutenant (2LT) Gabriel, who instruct us. They're from the 35th Batch. They went to OCS together.

Because commissioned officers outrank non-commissioned officers, Jackson and Gabriel are the two highest-ranking instructors we have. I think it strange that a 20-year-old who went to OCS can outrank a veteran warrant officer, but that's how it works.

A rotating group of sergeants, ranging from 3rd Sergeants to 1st Sergeants, assists our senior instructors.

HOOYAH

In NDU, trainees traditionally greet high-ranking officers as a class. We "hooyah" them. One man will shout the officer's title, dragging it out. The rest will echo, "Hooyah," followed by the title.

For example, whenever we encounter Master Chief NDU, someone will yell, "Maaaaaster, Chieeeeeef!" The class will yell back, "Hooyah, Maaaaaster Chieeeeeef!"

The officer will reply with a "Hooyah."

TUESDAY, MARCH 26

This morning brought our first tough run. We always run as a class, with the slowest at the front and the fastest at the rear. For the fitter guys, the runs are easy. We ran 4 km at a medium pace, which wasn't difficult—the hard part came afterwards.

We divided into groups based on our fitness levels and trained separately. I was part of the fastest group. We did interval training, running laps around a secondary parade square close to Block 29. Each lap is slightly less than 400 meters. Though we only ran four rounds, I was winded by the end.

My right hip joint, which I injured before enlisting, has been completely pain-free. As we were cooling down, though, I was hit by the worst pain to date—a brief searing needle. After that, it was fine for the rest of the day. I'm not sure what to make of it.

A PT session followed our run. We did 10 sets of 15 push-ups, three sets of 30 sit-ups, and four sets of 20 squats. It took place at the Grinder, a rectangle of smooth gray concrete under a high roof of corrugated metal.

After we finished the exercises, we were issued a mini-deal: instructors offered us a pull-up challenge.

One recruit from each section — eight in all — would compete against eight of our instructors and officers to see which team could do the most total pull-ups. If we won, the PT would end. If we lost, we'd continue on to squats.

Our eight guys went first. We yelled out each rep as a class, booming out the count and cheering until Cheng Kai eked out pull-up number 174 for our side, an impressive average of over 21 per recruit. We felt confident until the first two from their side, Instructor Shane and 2LT Gabriel, combined for 55.

After seven instructors had finished, their combined total was 170. Instructor Phillip easily banged out five more, and onwards to squats we went. Nevertheless, it was a fun bonding exercise. It also showed me that relative to the class, I'm definitely not as good at pull-ups as I thought I was. The best I can manage is 14 or 15.

After lunch, we worked with our rifles. We learned how to strip and assemble our weapons. We also covered immediate-action drills, which deal with how to handle issues while firing.

DEAR DIARY

As I write in my notebook, it's 2320. We just spent an hour and a half learning how to prepare our field packs for tomorrow's route march. We had to fold all of our clothes, arrange them in precise fashion, and pack them in Ziploc bags. It's simple, but it took forever to spread the information among 98 incredibly easily-distracted recruits.

WEDNESDAY, MARCH 27

Happily, thanks to Good Friday, we're booking out tomorrow night. Unhappily, this morning was the most painful to date. We had our first route march.

Though it was only 4 km, we wore our helmets, load-bearing vests (LBVs), and carried our rifles. Unused to the weight of the gear and not conditioned for marching, I found it mentally sapping and physically painful. My neck, traps and shoulders ached from the weight of the gear.

By the end of our hour-long plod around Sembawang Camp, my morale was flattened. I was drained. I wasn't getting any fitter, I was sore, and I didn't see the point of the march.

After the march was another stand-by bed. In response to another subpar performance, we carried out a change parade, repeatedly changing from our number four to our PT rig in a given time limit. If the entire class didn't make the timing, we were punished.

CHILLVILLE, POPULATION US

We're about to head to the cookhouse for dinner. Afterwards, we have no program; we'll be allowed to relax. This should be an easy night. I'm jotting this down at 1830, a pleasant change from scribbling in the dark after lights-out.

THE WATER PIPE

I spoke too soon. A few guys were assigned to pick up leaves from the field behind Block 29. For fun, they threw the leaves up in the air and tried to catch them. In the process of this exercise in maturity, Aneirin tripped over Block 29's main water pipe, breaking it. Water sprayed everywhere.

After laughing catastrophically, we realized we were screwed. The bathrooms and showers didn't work. We no longer had any water, and Instructor Shane was due to check on us in minutes. We decided to say Aneirin tripped over the pipe while collecting leaves, and hope that we wouldn't be confined.

Warrant Foo and Warrant Jaya arrived. We all thought we were going to be pumped. Amazingly, they promised we'll still be booking out tomorrow.

THURSDAY, MARCH 28

It's 1915 on a Thursday, and I'm home. I have a four-day weekend ahead and I'm stoked.

The training program for today was to tackle the SOC in the morning and then swim until book-out. The SOC is the Standard Obstacle Course, a timed evolution we must pass to complete Basic Military Training. Sembawang Camp doesn't have one, so we traveled to Nee Soon Camp.

As we prepared our gear for the SOC, Aneirin approached me with a smile and a book. Master Dennis had given him a book on leadership. It looked like a punched ticket to OCS. I congratulated him. On enlistment day, Aneirin and I joked about how great it would be if the two of us became officers. It looks like we're halfway there. I was surprised at how early this move was: we're only in the third week of BMT, and officer cadets are selected only after the Combat Diver Course. We have nearly six months to go.

Of course, I don't mean to imply that Aneirin and I are competing for places in OCS. Yes, we would both like to be officers, but we're not rivals. We're friends above all else. He's my best friend, actually. If he made OCS and I didn't, I would only be happy for him. If the reverse were true,

I'm sure he'd be happy for me.

A few minutes after Master Dennis spoke to Aneirin, Warrant Jaya beckoned to me. *Oh, shit,* I thought. As I jogged over, I tried to think of anything I might've done wrong.

"Sir?" I asked.

Warrant Jaya said, "Max, you've done a good job so far, and I'm not going to lie to you. We'll probably send you to OCS. There is something you must do as a leader, though, and that is get your hands dirty."

Surprised, thrilled and confused, I listened on.

"What I mean is that you must get a feel for the air down below. You must mix around, talk to many people, and know how they're feeling. Not only the high-flyers, but also the people you know who won't move up. To be a leader you must know how everybody feels, not only your friends. Are you going to sign on?"

"I'm considering it, sir, but I haven't yet decided," I stammered.

"Good," Warrant Jaya replied. "We hope you do. There is great opportunity for advancement in SAF. If they recognize potential in you, they will move you along quickly. You may wish to spend more time on the ground, but they might need decision-makers, so they may push you along faster than you'd like. So for now, what you need to do is get your hands dirty. Keep talking to people, get a sensing of the air, and stay in touch with everyone. Do you understand?"

"Yes, sir, thank you sir. I appreciate this very much."

"Good. Keep up the good work."

I was thrilled that it looked like Aneirin and I are going to become officers together. I looked around for him, excited to share the news, but Warrant Foo boomed, "39, muster now!" The news would have to wait. We scrambled into position for the SOC.

Master Dennis, Platoon 1's platoon sergeant, yelled for us to make some noise. "PLATOON ONE," we roared in unison. We were amped up.

The SOC is designed for three participants at a time. I was standing next to two of my section mates: my cabinmate Thomas, and a good guy named Harry. We pledged a friendly bet. The last one to finish would have to buy the winner a soya bean drink. We shook on it.

"Go," the instructor shouted, and we were off.

The SOC consists of 12 obstacles spread over a distance of 300 meters. It tests our speed, endurance, strength, and coordination. The challenges range from vaulting over walls to climbing a rope to crawling through a tunnel. After completing the obstacles, we turn around and sprint 300 meters back to the starting point.

Thomas, Harry and I ran hard. It was a good race. As a result, I am pleased to report that Thomas will soon be buying me a tasty soya bean beverage.

DROWN-PROOFING

After a canteen break and lunch, we moved to the Nee Soon Camp pool. Today, after several training sessions, the drown-proofing tests would begin.

Warrant Foo asked, "Are there any volunteers to go first?" Two guys quickly raised their hands. I thought, *Ah, shit. I wish I'd volunteered.*

"Good," Warrant Foo said. "And after them?"

My hand shot up.

Before I knew it, it was my turn. My pulse tightened as my hands and feet were bound behind my back. No goggles allowed. *Stay cool. Stay relaxed.*

I stood at the edge of the pool, nervous but ready, when Warrant Foo approached me. "*Ang moh,*" he said, "you got Diver's Gold, right?"

"Yes, sir."

"So, you are booking out early today, right?" This was a previous reward the instructors had offered: get Diver's Gold, book out early.

"Yes, sir."

"I tell you what. If you pass drown-proofing on your first try, I'll reward you with one extra day off. But if you don't pass, you won't book out early today."

I thought hard and quick. Drown-proofing was difficult. I didn't want to risk my early bookout.

"No, sir, I'm okay."

"No? You don't want the deal?"

Aneirin was standing nearby, casually eavesdropping. He

yelled, "Aw, come on, Max! Let's go, *ang moh*!"

Impulse flooded me. "Yes, sir! I'll do it. Sign me up, sir!"

"So you want the deal?"

"Yes, sir, yes I do."

"Okay then. Flynn, you're my witness. You heard that, right? Good luck, West."

I faced the pool and inhaled. *All right. Here we go.*

"Ready?" Warrant Foo clapped me on the back. "Go."

The drown-proofing test has six components. I jumped in and began the first: traveling. Traveling simply consists of swimming 50 meters. Normally, this would be simple, but with your hands and feet tied, it's tricky. I wriggled my way across the pool, undulating my body and kicking with my tied ankles. As MSG Martin taught us, it's all about the undulation. The test is untimed, so I moved slowly, conserving energy.

I finished without trouble. *One down, five to go.* Next up: flotations. To prove our survivability, we need to float for extended periods of time. I curled into a fetal position and floated near the surface, surfacing every few seconds for a breath. I needed to complete 15 breaths to move on. The trouble is that I'm negatively buoyant: even with a lungful of air, I sink quickly. Instead of floating, I had to repeatedly kick myself to the surface to breathe.

I was relieved to move on to bobbing, the third part of the test. For me, bobbing is the easiest portion of drown-proofing. We sink to the pool floor and launch ourselves back to the surface, completing 15 bobs at a depth of 3 meters. I relaxed, exhaling slowly, catching my breath.

Now, for the acrobatics. The next two stages were to complete a forward somersault and a backward somersault on the pool floor. Suddenly, I got nervous. I was worried about the somersaults. I would need to stay down for a long time. *What if I don't have enough air?*

Fortunately, there's no limit on bobbing—though it's mandatory to complete 15, we can do more if we want. I did three extra bobs before raising the nerve to do the front somersault, and then an extra two bobs before the backwards flip.

The final challenge was to use my mouth to pick up a mask from the pool floor. I sank once more and opened my eyes, peering through the liquid blur. I scanned the tiles. *There*—I spotted a black smudge. I bit down on the mask strap and shot to the surface. Mission success. It was over. I'd completed drown-proofing.

NOT SO FAST

Or, I thought I had. An NDU tradition, which they declined to tell us about before the test, is that after passing drown-proofing, we have to do 50 diamond push-ups. Then, still in the push-up position, we must shout "drown-proofing" as if we're hooyah-ing an officer. The class then replies, "Hooyah, drown-proofing." Only then is the evolution complete.

The first four of us to finish were Ivan Tan, Jay Yang, Tan Wei Qiang, and me. We did the 50 diamond push-ups together. I felt the burn, baby. I had to rest repeatedly, and the last five reps nearly killed me. When we finished, we hollered, "Drowwwwwn, prooooofing!" as loudly as we could.

Warrant Foo was standing nearby. He grinned. "Not loud enough," he said. Shaking in the push-up position, we screamed twice more. Finally, we were loud enough.

"Hooyah, drown-proofing," our batch called back.

Warrant Foo said, "Recover. Change up, and go for your canteen break."

"Yes, sir!" we yelled with glee.

Drown-proofing is a challenging trial, but one necessary to provide us with the confidence and maneuverability we need to operate efficiently underwater. Throughout the test, every trainee is closely watched by an instructor diving with scuba gear. Should we need oxygen, the instructor always has a spare regulator at the ready.

◦─══○───○══─◦

An hour later, the lesson was done. We gathered under a training shed to change into our number fours and leave. I couldn't wait to go home. The six of us with Diver's Gold were allowed to book out directly from Nee Soon Camp. The rest would return to NDU, tidy up their bunks, then leave. It should have been a smooth dismissal, but we didn't change into our number fours quickly enough. We took our time, laughing and joking around. Warrant Jaya noticed. When we were finally dressed and mustered, Warrant Jaya said, "Everyone — trunks. Two minutes."

We scrambled to meet the timing. After two minutes, I had barely made the transition. About a third of the class hadn't.

Warrant Jaya said, "Drop." We fell into the push-up position.

1st Sergeant (1SG) Khong shouted, "Those who are done changing, stay there. Those not done, move!"

The slow changers rushed to their clothes. I wished I could join them, especially when 1SG Khong began whacking us. He shouted "down" as we pushed out the reps, rapid-fire.

"DOWN-one-DOWN-two-DOWN-three-DOWN-four-DOWN…"

I thought we'd do 20 at most. Twenty push-ups came and went. So did 30, then 35, then 40. Most of us were in agony by this point, especially those who'd done the 50 diamond push-ups after drown-proofing. When we hit 45, I couldn't do a proper push-up anymore. I just collapsed to the ground, and used my knees to help push myself back up.

"Hold it there!" Instructor Khong yelled.

Only one member of 39, Keshav, hadn't finished

changing yet. Keshav bear-crawled to the front of the class. From there, Instructor Khong once again screamed, "DOWN!" and we resumed doing push-ups.

Keshav frantically said, "My boots — my boots are stuck."

1SG Zhong sneered, "Take your time, please, don't rush. It's alright. They're only your batch boys. No problem."

When Keshav finally got his boot off, we had pushed out 63 push-ups. My arms and chest felt like molten lead.

"Recover!" boomed the magic word. We scrambled to our feet.

Warrant Jaya said, "Change back into number four. Four minutes."

We scrambled into our clothes as quickly as we could. I reached for my number four top, but 2SG Jie Jun said to me quietly, "Put that on last, it's the easiest thing to do. Do your boots first." I thanked him and finished with just seconds remaining. However, not all of us made it. Those who had finished changing dropped again. We pushed out another 40 before everyone was done. To date, this was our worst punishment by far.

"Recover," said Warrant Jaya. We stood, panting, the acid slowly evaporating from our muscles. "Listen, I don't like being harsh. We don't like being so harsh. But if you don't take me seriously, if you don't take your training seriously, then we need to be. Do you understand?"

"Yes, sir!" we screamed.

The conversation shifted to a topic far more pleasant: booking out. Warrant Foo asked, "Where are Ivan and Max?"

We raised our hands.

"You both took the bet earlier, right? And you passed drown-proofing, right? So, you will have your day off. The rest will book in on Sunday night. You will book in on Monday night. Any questions?"

"No, sir!" I couldn't hold back my grin. An extra day off was an unimaginable luxury.

Master Raj implored us to stay safe and healthy on the weekend. He stressed that while in uniform, we are expected to behave properly. This means no swearing, no drinking, no smoking. We're also not allowed to fool around with our girlfriends in public. Or, as Master Raj put it, "No sexism."

Everyone with Diver's Gold was allowed to leave straight from Nee Soon, but Warrant Foo offered us a ride to the nearest MRT station. Elston Sam and I accepted the offer, while the rest chose to take the bus back to NDU.

Warrant Foo told us he and the other instructors needed to have a quick meeting, but he'd be back soon. Elston and I sat under the training shed and talked for 45 minutes. The conversation was easy and it was good to get to know him. He went to Anglo-Chinese Junior College, where he was captain of the squash team. He's been dating his girlfriend for nearly eight months.

Eventually, we wondered what was taking so long. Trying to get Warrant Foo to notice us without being too obvious, we walked to the bathroom, coughing along the way. When that didn't work, we resorted to PT, doing push-ups and pull-ups — loudly and reluctantly — until Warrant Foo finally strolled over.

On the way to the MRT station, Warrant Foo opened up to Elston and me. I thought he was the gruff old officer he seemed, but we learned he took his conduct seriously. He solemnly asked us if we thought he'd handled an earlier situation properly. It was the first time I've heard him speak softly. It was clear he was disappointed with himself for how he dealt with the issue. He dropped us off at Yishun MRT station and off we went.

FRIDAY, MARCH 29

I'm enjoying Good Friday, the beginning of my four-day weekend. To this point, I've been pretty self-centered, so I better introduce a few of my batch boys.

THE OTHER ANG MOH

Aneirin Flynn is the only other Eurasian in 39. We hang out at every meal and the instructors give us shit for always being together. He's smart, but, like me, dreams of an unorthodox career. I want to be a writer; he wants to be a sculptor.

"We'll make fine waiters one day," he concluded.

A FEW OF THE DIVER'S GOLD BOYS

Ivan Tan is the best runner in the class. A nice guy, he's acquaintances with everybody, but only close to a few.

Raymond Hoe is short but shredded. Fitness is his passion. He's generous with exercise tips and technique advice, especially in working with guys to improve their standing broad jump. One of the nicest guys, Raymond consistently volunteers to help out with the batch.

Shawn Ong is possibly the kindest person in the class. He volunteers at every opportunity. He's also the most soft-spoken. When he's forced to shout, as when we're numbering off for the class head count, his yelling is the volume of someone's normal speaking voice. It's funny when he contorts his face to scream and such a soft voice emerges.

Elston Sam was a guy I didn't know until we rode with Warrant Foo and booked out, but I like the dude. I think we'll be friends.

A FEW OTHERS

Helmethead's actual name is Aloysius, but we've called him nothing but "Helmethead" since the first week. Before our first stand-by bed by Master Dennis, Aloysius locked his personal cabinet. Unfortunately, when Master Dennis asked him to unlock it for inspection, Aloysius had forgotten the combination. Master Dennis bashed the lock open with Aloysius's helmet, and then made him wear the helmet for the rest of the day. He told Aloysius, "This is to keep your brains inside your head so you don't get any stupider." From then on, he's been Helmethead.

Jay Yang is reserved, but perhaps the most dependable guy in the batch. Only a sore knee hurting his SBJ prevented him from achieving Diver's Gold. Though we aren't too close, I'm certain that if I ever need anything, I'll be able to count on him.

I'll introduce more as we go on.

WEEK
FOUR

MONDAY, APRIL 1

The batch booked in last night, but lil old me strolled into Sembawang Camp at 2100, just in time to rejoin the boys for a field pack inspection by Instructors Zhong and Choi. Then followed our water parade, a nightly ritual in which we down a full liter of water.

The instructors ordered our bottles filled to the brim. Thinking they wouldn't check, I didn't bother filling my canteen fully.

"PT formation. Ten rows of 10. Spread out. Bottles up, caps off," 1SG Choi commanded. He and 1SG Zhong began inspecting bottles.

There are no words for the despair I felt at that moment. There I stood, too lazy to follow a direct order from our two strictest instructors. I found myself thinking, *idiot, idiot, idiot, idiot*, in a continuous loop. If he punished only me, that would be bad enough, but what if he whacked everyone for my mistake?

Instructor Choi approached me. I thought a silent goodbye to my family. He peered into my open bottle. It was dark; I had a fleeting optimism that the shadows would mask my laziness. But Choi raised his middle finger and dipped it into my bottle. The finger emerged dry. Choi raised an eyebrow at me. I managed, "S-sorry, instructor. It's not filled to the brim."

Choi paused, then nodded and walked away.

I was speechless. *What just happened?* Choi is perhaps our crankiest instructor. Two rows away, Zhong was screaming and punishing someone for not filling his bottle to the brim. Why was I let free?

Lights-out was at 2300.

TUESDAY, APRIL 2

VIRTUAL BULLETS

Yesterday, Platoon 1 completed the Individual Marksmanship Trainer, or IMT, a live-firing simulator for the SAR 21. Essentially, it's an arcade game. Everyone is required to complete the IMT before firing live rounds next week. We fire 32 shots from various positions at various targets in various settings. In Platoon 1, only two recruits notched a perfect 32/32 — Cheng Kai and Aneirin.

Though I'm in Platoon 1, since I had yesterday off, I joined Platoon 2 at the IMT today. Six dudes from Platoon 2 had a perfect score, but I wasn't one of them. I missed two of my first four shots, later finishing with a measly score of 29/32.

The IMT is at Nee Soon Camp. As we fired our simulated shots, Platoon 1 was swimming. We marched to the pool to join the swim, but the chlorine level in the Nee Soon pool was too high. We waited for 5-tonners, SAF transport vehicles, to carry us to nearby Khatib Camp.

This afternoon, we used fins for the first time. We swam turtleback and sidestroke. Turtleback is a face-up flutter kick — picture backstroke without any arm movement. Sidestroke is a tactical swimming style. It's efficient and causes little disturbance on the surface, making it tough for the enemy to spot us. Shooting along in fins was wonderful, but my quadriceps and hamstrings were on fire.

We swam all afternoon. As we changed into our number fours and prepared to return to NDU, we were dropped again for not changing quickly enough.

Aneirin muttered, "Why can't these guys just put their clothes on faster?" I wish I knew.

* * *

After dinner, our final evolution was a stand-by bed. We had a generous 90 minutes to prepare. 3SG Phillip was the duty instructor. However, nearly all of us were confident that there would be no inspection. We figured Instructor Phillip would let us off easy for the night. My cabinmates and I were tired, so instead of cleaning, we took a nap.

Eighty-five minutes later, we were rudely awakened. "39, muster down in five minutes! 39, muster down in zero-five minutes! Stand-by bed in zero-five minutes!" Our current IC, Shiva, was shouting his brains out.

"Oh, shit," I said. "Guys, wake up." Benjamin, Noel and Thomas sat up. We sat in silence, staring at our inexcusably dirty, messy room.

We leapt out of bed and scrambled. From what it sounded like, everyone else was doing the same.

By the time all 98 of us had mustered, we were ridiculously late. Instructor Phillip snapped for Shiva to follow him, and began his inspection. He was angry.

Phillip said he had counted over 20 mistakes. He offered us a choice: spend 30 minutes cleaning and then stand for a re-inspection, or select Option B. He would not reveal what Option B was.

We hate cleaning. Option B was the unanimous choice.

OPTION B

Option B was bodybuilders, an eight-count exercise similar to a burpee. Each bodybuilder includes a push-up and a jump squat. We had to do 50 as a class.

We began pushing them out. When we reached 20, Helmethead was struggling. He couldn't keep up. Phillip shouted, "If you can't do them together, we'll start from zero."

Shiva called Helmethead to the front so he could set the pace. Now, we did them agonizingly slowly, pausing between each component of the bodybuilder. We croaked out five more before Instructor Phillip halted us again. He said, "39, you're doing them so slowly, I'm going to fall asleep. I'll take pity on you. Since you're batch 39, all I need from you is 39 bodybuilders."

We cheered. If punishment was tied to batch number, I didn't envy future classes. Batch 500 will be in for some shit.

We finished the bodybuilders and mustered for our water parade. Then, the unthinkable—Phillip noticed someone was wearing blue shorts.

In the SAF, and I imagine in every military, standardization is a religion. We wear the same things, eat and sleep at the same times, and do all our evolutions together. We were supposed to gather in blue T-shirts with black shorts. But Harry, my section mate, had worn blue shorts.

"Drop!" shouted Phillip. I steeled myself. Push-ups are bad, but holding in the push-up position is even worse. Push-ups are a familiar pain, one that dissolves quickly. Even when we did over 60 last week, it was over within a couple of minutes. But remaining in the push-up position is a slowly accelerating burn, layers of soreness rising in the arms and shoulders. It's prolonged. It's awful.

After a minute, we recovered and drank up. Phillip, disappointed, dismissed us.

SPIRITUAL AWAKENINGS

There was some positivity to end the night. Everyone has been shy about revealing their birthday. I think this is because it's tradition to be pranked on your day of birth. But today was Oscar's birthday.

Oscar is great. He's the shortest (157 cm), lightest (49 kg), and least fit member of our batch (2.4 km time of 14:50). He is, however, universally liked. Oscar is friendly and tries his best. His chances of successfully passing out as a diver are small, but we want him to make it. A few hours after Phillip released us for the night, 3SG Wayne walked up to Block 29 with a small box. It held a birthday cupcake for Oscar.

The news flew through Block 29. Soon, everyone was gathered in the hallway outside, laughing and cheering as Oscar went downstairs to receive his treat. We sang happy birthday to him as a batch, and then watched as Phillip made Oscar do push-ups for every bite of cake.

Instructor Phillip yelled, "See? This is class spirit! I want to see more of this, not that bullshit from before. Is that understood?"

"Yes, instructor!" we shouted back as one.

EARNING OUR PAY

Throughout today, I think we were dropped more than ever before. We were pumped twice during the IMT, several times throughout the swimming evolution, and then during and after the stand-by bed.

Hey, this NS allowance doesn't earn itself.

WEDNESDAY, APRIL 3

We woke up early to draw our rifles from the armory. We boarded buses for Nee Soon Camp at 0645. I'm still adjusting to the lack of sleep, but it's getting better. I'm not as forgetful as I was in the first weeks.

This morning was the first time we tackled the Standard Obstacle Course with our rifles. It was also the first time we were timed. The SOC is a veto. To graduate from BMT, we must complete the course within 4 minutes and 30 seconds, wearing our helmets, load-bearing vests with a filled 3-liter canteen, and rifles.

I like the SOC and I'm competitive. I ran my best, but as I crossed the finish line, I noticed nobody was recording times.

I found Aneirin, who'd gone before me. "Man," I gasped, "I don't think they timed us!"

After everyone was done, Warrant Jaya said, "39, I have good news. That was a practice run. You will have a 15-minute break, then we will do it again. This time, we will take your timings."

That really pissed me off. *That was a practice run?* I was basically being punished for trying my best. I had run hard, so I was tired. On the other hand, those who slacked earlier now had more energy left for the actual trial.

As I waited at the starting line, Instructor Khong saw me. He asked, "Max, what's your goal timing?"

I had no idea. The passing time was 4:30, so I offered, "Four minutes?"

Khong scoffed. "An old man like me can do it in 3:30."

Surprised, I said, "You can do 3:30?" He nodded. "All right then, I'll beat you," I said.

"Hey, you sure you can do it? I don't think so. You're not fit enough." The guys around me ooh-ed, recognizing the challenge. I knew Khong was baiting me, but it still worked. I got angry. *I'm gonna get 3:30.*

"Go," said the PTI, and I sprinted. I raced through the obstacles, blazing through each one. I finished the final obstacle well ahead of my cabinmates, who had started at the same time, but I still had a 300-meter sprint to go. 300 meters doesn't sound long, but when you're wearing a load-bearing vest, helmet, rifle and combat boots, and you've just navigated 10 obstacles, it's far enough. Especially when you just did the course 20 minutes earlier. Panting, I took off on the final stretch.

As I neared the finish line, I was exhausted. I willed my legs to sprint, but felt like I was only moving at a jog. I finished and looked to Khong for my time.

"Good," he said.

We swam in the afternoon, which was a godsend. We did a few relay races, but were otherwise able to relax and cool off in the water.

We never found out our times.

THE IMPORTANCE OF BEING ATTENTIVE

As we boarded the bus back to NDU, I passed Instructor Wayne. He grinned at me for no apparent reason. Confused, I smiled back, and sat next to Shiva. Then, Instructor Wayne shouted, "Magazine check!"

Shiva and I ran our hands to the bottom of our rifles, feeling for our mags. We froze. Our batch boys tapped their magazines, signaling they had theirs. But I felt nothing. I looked at Shiva, who appeared to be just as shocked as I was.

Instructor Wayne smiled at us and held up two magazines. "Looking for these?" he asked. 3SG Thaddeus and 3SG Russell burst into laughter. They agreed that a suitable payment would consist of 39 bodybuilders. I was relieved; I was expecting worse. Then, Instructor Wayne said, "No, Max, you and Shiva won't be doing the bodybuilders. The class will. You two will stand at the front and watch."

"Wait, instructor, no!" I protested. This happens frequently. When someone makes a mistake and the class is punished, the mistake-maker is often exempt— he'll stand and watch the rest suffer for his blunder. I've realized that it's far less painful to be punished yourself. When you cause the batch to suffer, they resent you. For instance, though Keshav and I are good friends, when we had to do 63 push-ups because he couldn't take his boot off, I had hated him for it.

Shiva and I managed to convince the instructors to

allow us to perform the bodybuilders ourselves, with the added embarrassment that we would continuously chant, "I will not lose my magazine."

Relieved, I thanked Instructor Wayne. He cackled, "I love this job! We punish them and they thank us."

⇥⇤ ⸺ ⇥⇤

Following dinner, it was a quiet night at 29. The cookhouse had run out of Milo powder, spoiling our night snack. The snack is usually Milo and a bun, muffin or cookie.

The best part of night snack is the labels. Each packet is stamped with the item and its flavor, but the flavor is always listed *after* the food. For instance, instead of a banana muffin, we have "muffin banana." Other favorites are "cake chocolate," "bun raisin," and "tart pineapple."

CAUTION: PAIN AHEAD

My shoulders were sore from carrying around our LBVs and rifles today. At the end of BMT, though, we'll have to endure far worse. Before we pass out, we'll complete a 24-km route march, wearing our full battle order—helmet, LBV, rifle and field pack. In all, the load is over 15 kg. Those 24 kilometers are going to suck.

THURSDAY, APRIL 4

THE LONGEST FOUR

This morning, we did a 4-km route march in full battle order. The instructors say we'll progressively train up to the 24 km, gradually increasing the distance. But these four kilometers were already an eternity. The worst ache is in the shoulders. At one point, I even considered "accidentally" tripping, feigning an injury to escape the rest of the march. I couldn't actually bring myself to *chao keng*, but though it was only 4 km, that march was one of the tougher evolutions we've done. I can't imagine marching 24 km.

We sang almost constantly. We always sing in the same fashion, with one man shouting a line and the rest of the class echoing. At first, I was reluctant to join in. I thought it made more sense to save my breath. Near the end of the march, though, I decided to sing along with the class. To my shock and wonder, I found the march instantly became easier. I felt uplifted, part of a greater whole. Singing brought my mind outside my body — no longer was I turning inwards, focusing on my pain.

We left our field packs at Block 29 and mustered outside the cookhouse for our pull-ups. Since the second week, we've been doing pull-ups before every lunch and dinner in order to "earn" the meal. We often wear our LBVs. The stronger ones do 12 pull-ups and the weaker ones do eight. Those who can't do eight on their own are assisted by the next man in line.

Next were more rifle lessons. We headed to the NDU multi-purpose hall, located directly above the cookhouse. We split into sections. Our section leader, 3SG Shane, taught us how to load and unload our magazines. We practiced with blank rounds. There was a brief timed test to make sure we can load rounds quickly enough.

CHUCKLE CITY

It was pouring outside, so we remained in the hall and had a "comedy club" session. At the far end of the hall is a large stage. To keep us occupied, one recruit at a time went onstage and told a joke. If it was funny, he could rejoin the class. If not...

As IC, Shiva went first. His punch line was punchless. Instructor Russell immediately said, "Drop."

Aneirin volunteered himself onstage. He asked, "Why is Santa's sack so big?"

"Why?" we asked.

"Because he only comes once a year."

We laughed. Instructor Russell let him escape.

FRIDAY, APRIL 5

Our training schedule for the week is posted on a bulletin board at Block 29. This morning, a 4-km run was scheduled. When we awoke to streaks of lightning, I rejoiced. I didn't feel like running. I prayed for rain. The bursts of lightning continued through breakfast, but the clouds held. I kept praying, but by 0800, it was hot and dry, and we were mustering in PT attire.

Instead of doing this run as a class, we split into three groups, optimistically classified as fast, faster, and fastest. Those with a 2.4-km run time below 10 minutes were placed in the fastest group, led by Instructors Thaddeus and Russell.

About halfway through, the slowest member of the fastest group, the soft-spoken Danny Tan, began to falter. Aneirin and I dropped back from the main group and ran alongside him, trying to spur him on. Eventually, Instructor Thaddeus ran with Danny and a few others who were struggling, while Instructor Russell fronted the pack.

On the run's final stretch, Instructor Russell gradually accelerated into a full sprint. We flew through the NDU gate and finally slowed, jogging at a recovery pace until we found our breath. We finished the run in just over

17 minutes. Russell told us we had actually run a bit farther than 4 km, which was welcome news.

———❦———

During lunch, the batch was buzzing. We all knew it was book-out day. We were excited. Our goal was simply to not screw up.

We marched to the auditorium. Next week, we have live firing with the SAR 21, the standard-issue weapon of the SAF. Designed and produced in Singapore, the weapon's name is short for "Singapore Assault Rifle — 21st Century." I've never fired a gun before. Shooting real bullets will be awesome.

Master Raj delivered a safety briefing. He's still the nicest instructor we have. He's never dropped us or even raised his voice.

I was sitting next to Elston. I whispered, "Dude, Master Raj is so nice. I can't believe it."

"You know what I heard?" Elston said. "Master Raj used to be the most sadistic instructor in NDU. He used to be damn fierce."

"No way. Really?"

"Yeah, really, really!"

I didn't believe Elston, but at the end of the presentation, Master Raj said something that convinced me. While discussing trigger discipline, Master Raj's face took on an entirely new expression. His gentle eyes turned metallic and his soft voice hardened into iron. "If anyone — anyone — doesn't follow the safety procedures, I will punish you. Severely. Believe me. You may think I'm nice, because I've been nice so far, but you do not want to test me."

Murmurs of surprise wafted through the auditorium. Elston nudged me. "See?"

With that, the briefing ended. It was around 1630, too early for dinner, so we hung around. Without warning, the Commanding Officer of Dive School, a Lieutenant Colonel (LTC), walked in. Shiva led us in a hooyah, yelling, "CO, Diiiiiive Schoooool!"

"Hooyah, CO Dive School!" we roared.

"Hooyah," he answered back.

Then, CO released a bombshell. He said, "I have some news for you all. As you know, a few members of your batch have sore eye. We were hoping it would not spread, but that does not appear to be the case. In light of this, as a preventative measure to protect your health, we have made the decision to give you two weeks off from training. Instead of booking in this Sunday night, you will book in in two weeks' time."

Shock and glee streaked through the batch. *Two weeks off? Is he serious?*

CO continued, "This may come as a surprise to many of you, but we feel it is in our best interests. Your health and welfare are our priority. When you return, however, should the sore eye issues persist, we will simply have to accept them and push through. Are there any questions?"

There was a pause of roughly five seconds, and then a wave of hands darted up. CO laughed. Most were happy about the forced vacation, but Elston said to me, "I feel like two weeks is actually too long."

I completely agreed. I'm nearly used to life in NDU. Two weeks off will shatter the rhythm. When we return, it'll feel like enlistment day again.

Keshav agreed with us as well. He said, "You should ask if we can have one week off instead."

"No la," Elston responded, "you do it!"

Following CO's departure, Warrant Jaya and Warrant Foo arrived. They had just heard the news as well. They briefed us on technicalities: our total service time will not be extended, though our BMT will be. Our pay won't be affected, either.

Instructors Khong and Zhong entered. They brought with them bad news: our first week back will be field camp.

Field camp, or Individual Field Craft (IFC) training, is one week in the jungle. It's rumored to be the worst week of BMT. A week in the jungle without a shower, a real meal, or a solid night's sleep. And it will be coming just as we've reacclimatized to civilian life.

We were also instructed to get flat-top haircuts. Warrant Jaya explained that in the 1970s and '80s, the flat top was the mark of the naval diver. "Everywhere we went, people saw the flat top and knew we were divers," he said. NDU wants to resurrect this tradition. Saucy.

We returned to our bunks to pack our things. We had nothing more to do, but evidently, it was too early for us to book out. We stayed in our cabins and hung out. At last, at 2100, Instructor Khong summoned us downstairs.

We walked out of Sembawang Camp and were free. So long, NDU — it'll be a while.

The 39th Batch stands at attention.

The author (center) flashes a rare smile during a nighttime run.

The author (foreground) leads the batch through a set of push-ups.

The author (left) and fellow ang moh *Aneirin Flynn enjoy a dip in the NDU dive pool.*

THE BREAK

SATURDAY, APRIL 6 – SATURDAY, APRIL 20

Two whole weeks off from camp. Can you believe it?

I've been sleeping 10 hours a day, but I still feel tired in the mornings. Going back to the usual wake-up at 0515 isn't going to be fun. I haven't been training intensely, but I've gone for a couple of runs and swims. I hope it's been enough to maintain my fitness level. I guess I'll find out when we get back.

NDU feels a long way away, almost like a dream. I feel myself falling back into my old life. Everything is almost exactly as it was before I enlisted.

But I'll be back soon enough.

WEEK
FIVE

SUNDAY, APRIL 21

We booked in tonight. I thought it would be strange to return to NDU, but it felt like we'd never left. We leave for field camp on Tuesday morning and return on Friday.

My birthday is April 22, which astute readers will notice is tomorrow. Aneirin and Keshav insisted I stay up till midnight, so we hung out in Keshav's room with Elston, Brendan Conceicao, Rahul, and Dean. I don't know Dean well, but I'm good friends with Brendan and Rahul.

Because it's traditional to be pranked on your birthday, I deliberately sat closest to the door, ready to make a dash. At 2350, everyone but Brendan and Dean left for "the bathroom." Figuring they were preparing something, I waited 30 seconds and then dashed for the door — but it was locked from the outside. I was stuck.

At exactly midnight, I heard the door unlock. I braced myself. But instead of a prank, they entered with a cake and a card, singing happy birthday.

Aneirin made a card "from the NDU boys," featuring a silver frogman silhouetted on a red background. Elston had smuggled in a piece of cheesecake and candles. I was touched. Guys from neighboring rooms heard the commotion and came outside to shake my hand.

They began discussing my birthday dare, what I'd have to do to "earn" the cake. The ideas included having me run around 29 naked, but they settled on me having to ask Master Dennis for a hug tomorrow.

It was the best birthday I've ever had. I've never liked cheesecake, but that slice was damn good.

MONDAY, APRIL 22

The morning of my birthday began with a 4-km run. Thanks for the present, NDU. The pace was moderate, but it was more tiring than I expected. Interval training followed, as we again ran laps around the secondary parade square near Block 29. Each lap is roughly 350 meters. We had to complete each round in 1:15 or less.

"You will be doing eight intervals," Warrant Jaya said. This terrified us. When we stopped after just three, the air was light with relief.

We proceeded to the NDU football field, a grassy rectangle with a concrete grandstand on one side. The instructors demonstrated contact drills, which we'll execute during field camp. There are four general ones: contact front, contact rear, contact left, and contact right.

⋅⇥▆▇ ⎯⎯ ▇▆⇤⋅

We returned to the grandstand after lunch. Master Dennis appeared before us, holding a white box. He yelled, "Max! Where's Max? Get Max down here."

Uh oh. I walked to him uneasily, expecting punishment. Instead, he opened the box. He'd bought a chocolate cake for me. 39 sang me happy birthday as a class. I was thoroughly embarrassed and wholly touched. Master Dennis dropped me for 39 push-ups. The class counted out each rep.

Aneirin and I cut the cake into eight pieces and gave them to the section ICs to distribute. I thanked Master Dennis. When I looked up, I saw Keshav and Elston mouthing, "Hug him, hug him!"

Damn it. They had brought me cheesecake, after all. I had no choice.

I tapped Master Dennis on the shoulder and asked, "Sir, could I have a hug?"

"Screw you!" he shouted, to the class's delight.

I broke off a small piece of cake for myself. I was lifting it to my mouth when someone tapped me on the shoulder. I turned and saw a brief flash of brown before I was stumbling backwards and 39 was laughing their asses off.

Master Dennis had slapped me in the face with a huge piece of cake. I tried to rub it off before the class could see, but there was a centimeter-thick layer of chocolate smeared on my face. All I could do was embrace it. I turned to face the grandstand and shrugged, as the class continued laughing. Rahul shouted, "Now you look like me!" I went through five minutes and roughly a thousand napkins, but I eventually cleaned the cake off my face and hair.

It was definitely the best birthday I've had.

We spent the afternoon loading gear into tonners for

field camp. After dinner, we packed our field packs and then were allowed to rest. I went for a shower.

I should've seen it coming.

THE HAPPIEST OF BIRTHDAYS

Halfway through my shower, the door banged open, and raucous laughter flowed into the bathroom. I turned to see a naked Aneirin, Keshav, Rahul and Elston swarming my shower stall. Shouting, they pushed me around and group-hugged me, cracking up the entire time. Aneirin yelled, "Happy birthday, bitch!"

They were lathered in soap, too slippery to push away. They finally left when I grabbed my towel and started whacking them with it.

It was a hilarious, slightly traumatic experience. It's fair to say my first time showering with someone wasn't exactly what I had in mind.

We need to be mustered at 0430 tomorrow morning to have breakfast, load our food rations and water into tonners, and leave for Mandai Forest. See you in a few, NDU.

TUESDAY, APRIL 23

FIELD CAMP

We mustered at 0430, minds cloudy with sleep. In a haze, we ate breakfast, loaded stores, and climbed inside tonners for the 45-minute ride to the Mandai jungle. En route, we managed to convince our section IC, Jared, that Mandai is in Malaysia and that we had a five-hour drive ahead.

I nodded off. When I awoke, my tonner was growling through the jungle, along a dirt path carved through the trees. We rumbled to a stop in a gravelly clearing. There was a large training shed which would serve as HQ for the week.

We marched to our campsite about half a kilometer away. It was an area of thin jungle, about 100 m by 30 m. There was enough space to dig our shellscrapes and erect our *bashas* between the trees. A basha is a crude tent fashioned from a ground sheet and some sturdy cord. One basha can house two people.

Master Dennis lectured us about basic field and noise discipline. We are to treat field camp as seriously as if we're truly hiding in enemy territory. We are not to speak, communicating instead with hand signals. While resting, we never stand; everyone takes a knee. (In full battle

order, high-kneeling rapidly becomes uncomfortable.)
At night, we don't use our flashlights unless absolutely
necessary, and even then, only with a red lens.

Next was a lesson on shellscrapes. They're shallow
indentations dug out from the ground, long enough
to lie in, deep enough to provide cover. We'll use them
for nighttime sentry duty. Master Martin instructed us
to plan the arrangement of our shellscrapes and within
the campsite. The bashas were to be centered, with our
shellscrapes strategically placed to defend our position.

Our ICs for this week are Vincent and Zeng Hou.
They might be the two tallest guys in the batch. While
Vincent is a bit of a joker, both have been relatively quiet
so far. Everyone was surprised when they volunteered to
be ICs.

Aneirin told me that after they volunteered, Warrant
Foo pulled him aside and asked, "Flynch, who the hell
are these guys? Will they be good ICs?" (Warrant Foo has
trouble pronouncing "Flynn." He always says "Flynch.")

"I'm not sure," Aneirin said.

"Fine, Flynch," Warrant Foo said. "I'll give them one
day in the field. If they screw it up, they'll be gone."

Well, neither one of them is off to a good start. Zeng
Hou is in my section, and while he's friendly with us, he
is awfully shy around everyone else. Vincent has trampled
over Zeng Hou at every turn, dismissing each of his
suggestions. Worse, Vincent himself makes questionable
calls.

We're to dig two shellscrapes per section, so we'll have 16 to place around our rectangular campsite. One length of our site runs parallel to the road, while the other faces the deep jungle, so Vincent chose to focus on guarding the roadside. This was fine, except he skewed our resources dramatically: of our 16 shellscrapes, he placed 12 facing the road, with just four left to defend the rest of our perimeter. Over multiple objections, he voiced this plan to Master Martin. Master Martin was as puzzled as the rest of us. Vincent relented and we distributed the shellscrapes more evenly.

We headed to the training shed to collect our combat rations. A day's worth includes two meal packets, one dessert packet, and an accessory pack with biscuits, sweets and powdered drinks. The food comes in dark green plastic packets. They're soft and mushy, like mashed potatoes. There are a variety of choices. We picked packets at random. I drew chicken with glutinous rice and mashed potatoes with chicken sausage. Deciding to save the tastier-sounding packet for dinner, I tore open the gelatinous rice. It was actually pretty good.

DIG IN

Carving out our shellscrape proved challenging. I was digging with Yew Ming and Scott, two of my section mates. Yew Ming is lean and ripped. Scott is a chubby recruit with nipple hairs long enough to touch his armpit. They're disgusting. After 30 minutes, we noticed our budding shellscrape was about one and a half times as wide as necessary. Instead of fixing the problem, I shrugged and said, "More space for us to lie in, right?"

That decision was not my best. We kept hacking at the dirt, but since we were digging at a far larger surface area, our progress was slow. Instructor Khong strolled by and said, "Wah, you guys digging a jacuzzi or what?"

Forty-five minutes later, our shellscrape looked like shit. Instructor Phillip walked over and laughed at us. He said, "With Max digging, I thought this one would be good, but I guess not."

Another hour later, we were drenched, blistered and sore, but we were done. It looked messy, but I thought it would suffice. We were wrong. Master Dennis looked it over and barked, "Max! You want to go to OCS, don't you? Is this the best you can do?"

Well, screw me, I thought. "No, sir!"

We spent another 30 minutes undoing the extra work we did before, filling in our extra-wide shellscrape until it was a normal size. Finally, Phillip gave us a nod of approval. It was done.

Scott, Yew Ming and I were soaked with sweat. Adam Ong walked by and said, "You guys take a shower or something?"

QUEEN OF THE BATTLEFIELD

By 1930, the night was asphalt black. We weren't allowed to use our torchlights, either. Lying in our basha, my buddy Thomas and I shoveled down our combat rations. The delicious-sounding mashed potatoes with sausage, which I had saved for tonight, turned out to be absolutely disgusting.

We lay in the blackness.

I had a sudden urge for company. I said, "Hey, wanna go find some people and hang out?"

"Sure," Thomas said. We walked to the far corner of the campsite, where we found Blake, Thanaraj, Siu Hoi, Sean Teh, Shiva, and Elston relaxing, talking quietly. As we chatted and swapped stories, our field discipline slipped. We went from whispering to speaking softly to speaking normally to laughing out loud. Somehow or other — I really have no idea how — we ended up singing "Bohemian Rhapsody" by Queen, belting out a terrifically tuneless rendition. Later, we would learn that Instructor Khong yelled for everyone to shut up, and the batch began yelling the same. The eight of us didn't hear anything, though. We kept singing.

Too late — we saw the blue lightstick too late. Only the instructors had blue lightsticks. But we were too focused on nailing the high-pitched "Galileo" that we didn't notice. Instructor Khong was there, and he was pissed. He tore into us. Our group of merry singers dispersed.

I thought that would be the end of it. Ten minutes later, 39 was called to muster at the training shed. Master Dennis was furious.

First, he called about 15 of us out from the group — Aneirin, Keshav, Rahul, Elston, Ivan, me, and others. He said that he saw potential in us, so he was going to talk to us first. It started out as a soft-spoken "I'm disappointed in you" lecture, but Master Dennis progressively became angrier until he was screaming. Our field discipline had fallen to levels inexcusable. He shouted, "And one group was singing! Singing in the field. What the hell kind of military do you think this is? Don't worry. You will pay." Shame filled my chest, thick and hot.

We mustered behind the training shed in open order, with a gap between each row. Master Dennis said, "Two minutes. Everything off." When we stood in our trunks, he said, "Two minutes. Everything on."

The change parade lasted half an hour. Stripping and donning our number fours repeatedly isn't physically tiring, but it's stressful. The time limit descends like an explosive's countdown. We're all aware that one or two slip-ups and we'll be too slow. Nobody wants to be singled out for failing the timing.

I was on the far left, in the second-to-last row. A few meters away, I saw Instructors Thaddeus and Russell using a torchlight to chase around the biggest centipede I've seen in my life. It was a foot long and thick as the handle of a baseball bat. It was absolutely terrifying.

Eventually, those who made the timing were "secured," or finished with the evolution, and were allowed to rest at the training shed. I finished and joined the others. From a few rows back, Keshav hissed, "Max! That wasn't too bad, right?"

I turned my head to whisper for Keshav to shut up, but Master Dennis saw me. "Max! Keshav! You think this is a time for talking? Drop!"

We stepped out of the shed and assumed the push-up position. I yelled, "Permission to carry on, sir!" No reply. I repeated the question, louder this time. Still no answer.

Instructor Wayne was nearby. I asked him if we should just proceed with the push-ups, but he motioned for me to shout again.

"Keshav, we should shout together," I said.

"Nah man, you try, you try," he said. I shouted again, "Permission to carry on, sir!"

Master Melvin whirled. He walked over to us and roared, "Keshav! How come Max is the only one who can shout for permission? You don't have a mouth, is it?"

Together, we shouted, "Permission to carry on, sir!"

"Carry on," Master Dennis said.

Keshav and I pushed out 40 before rejoining the class. Getting punished is painful, but Master Dennis's disappointment hurt just as much. He was teary-eyed at one point. I felt like we betrayed him.

When the change parade was done, we mustered in preparation to return to our campsite. Master Dennis found me. He said, "Are you proud of yourself?"

"No, sir," I said quietly.

Our campsite was as silent as a graveyard. Section ICs whispered out that night's schedule for sentry duty and we went to bed.

Every buddy pair had to do one hour of sentry duty at some point in the night. Thomas and I drew the 3 am shift, because of course we did. It was tough to stay awake, but I brought some crackers to nibble on, and that helped. Surprisingly, the time passed quickly.

Shortly after our shift ended, we were lying in our basha. I was mired in that muddy slipstream between wakefulness and sleep. Our basha was on a slope, and I had the sudden sensation of sliding down a hill. I flung my elbows outwards and dug them into the ground to anchor myself. After a few seconds, I shook myself into consciousness and realized two things — one, I wasn't actually sliding anywhere; and two, I'd just elbowed a very confused Thomas in the chest.

"Sorry, dude," I mumbled, and went back to sleep.

WEDNESDAY, APRIL 24

Our second day in the field brought lessons on cover and concealment. This began with camouflage cream, or "camo," and progressed to lessons on how to hide in the jungle. We played a game — one section would take cover and another would try to spot them.

I managed to royally mess up my camo cream. We apply a base coat of green to the neck, face and ears, and then add three black diagonal stripes across the face. After Shiva helped me put on my black stripes, someone mentioned my camo cream was too thick in one spot. After I smoothed it out, I decided to spread around all of the camo on my face to ensure it was evenly coated. I did not, however, realize that I would be spreading the black stripes all over my face.

When I finished, I turned around. Immediately, four or five of my section mates pointed and laughed at me. Only after they laughed for a solid five minutes did I figure out what I'd done. Needless to say, I had to reapply my camo. By the time it was fixed, I felt like I had a layer cake plastered onto my face.

We moved on to the contact drills. The procedure is simple but difficult to execute smoothly. At the first sign of enemy fire, everyone drops into the prone position, facing the threat. Crawling, we arrange ourselves into a straight line called the firing line. We then leapfrog towards the enemy: half of the line runs forward while the other half provides suppressing fire. While leapfrogging, our objectives are to spread out, take cover, and keep the firing line intact.

The process continues until we're within 50 meters of the enemy or they are defeated, at which point we charge. It's not really a charge — we stand up and walk slowly towards the enemy, firing as we go. Throughout the entire drill, the section IC gives the commands.

The contact drills sound basic, and they are, but they proved hard to coordinate. The most common issues were a crooked firing line and improper orders given by the section IC.

⊷═◉──────◎═⊶

When we returned to the training shed for lunch, Warrant Foo announced that two recruits' rifles had been stolen. One belonged to Mark Liang. The other belonged to Aneirin. As a result, Warrant Foo continued, their punishment was that they would take over as ICs for the remainder of field camp.

"What happened?" I asked Aneirin.

He said, "I just turned my back on it for a second to polish my boots. When I looked back, Zhong was running away with my rifle. I chased after him and caught up, but he wouldn't give it back."

He was upset. He hadn't even done anything stupid. I'd be upset too. Hell, almost all of us have turned our backs on our weapon at some point. Aneirin just got unlucky.

It rained in the afternoon. Because of the lightning risk, we couldn't continue with our scheduled drills. All we could do was sit in the training shed and wait. An hour into the break, Warrant Foo said, "Everyone, listen up. Who here wants to be an officer?"

A large number rose, including me. I quickly scanned the shed. I counted 29 guys standing.

"All sit," Warrant Foo said. "Now, of you, who has Diver's Gold and also passed drown-proofing?"

Ivan, Elston, Peter Wong and I stood up.

Warrant Foo said, "Currently, these four have the highest chances of going to OCS. The rest of you, if you want to go, have some catching up to do. Unfortunately, I can only send *two* of you to OCS."

The class gasped. There would only be two officers from 39?

"The four of you, come up here," Warrant Foo continued. We sat down at the front of the shed before him, Warrant Jaya and Master Raj.

"We will now interview you about going to OCS. Max, you first."

I'm going first? Screw me. I took a deep breath and tried to stay calm.

Instead of asking me about myself, the instructors asked about Aneirin. I realized that they wanted to send Aneirin to OCS, but because he hadn't gotten Diver's Gold, they had to get information about him from me. I told them I thought Aneirin was an exceptionally strong candidate. Warrant Foo said, "I think you're right about Flynch. He's solid. Right now, though, he's lost his rifle. He's the lowest of the low. This is the lowest he'll be in his life. You said he rebounds quickly from mistakes. Well, we'll see."

The focus of the questions switched to me. Warrant Foo started with an easy one. "Why do you want to go to OCS?"

I said, "As I'm half-American, I'm thankful for this opportunity to do National Service, because if my parents were based in America, I wouldn't have this chance. As such, I'd like to benefit from NS as much as I can. I believe that I would be able to learn a lot at OCS."

They seemed satisfied. "Now, Max," Warrant Foo asked, "Give me three qualities about yourself that you believe would make you a good officer."

Earlier, he had asked me the same question about Aneirin. He continued, "And you must think of three different qualities than the ones you said for Flynch."

I saw him start to smile, so I laughed. This was a tough question. I struggled to phrase my bullshit modestly.

"Well, sir, um, I've tried my best on all the fitness tests so far. The IPPT, the drown-proofing, the Cat 1 swimming test, and I believe I've performed quite well so far."

"Yes, your results have been outstanding."

"I'm also able to think quickly on my feet and adapt to new situations. I believe I will be able to make decisions quickly — the right decisions, that is, sir."

Warrant Foo grinned and said, "Okay, so that's two. What's your third?"

I had nothing. I literally had no idea what to say, but I had to say *something*. "Umm, well, my third and final reason is that I believe — I know that I would enjoy being an officer. I wouldn't look at the responsibilities as a burden, but I would try to take pleasure in leading my men. Even the harder parts, like disciplining them, I wouldn't see as a burden, but I would rather try to enjoy the experience as a whole."

I had no idea what I was talking about. I didn't even know if that made sense. (In hindsight, it didn't.) Thankfully, Warrant Foo moved on. "Anything from you, Warrant Jaya?"

Warrant Jaya crossed his legs and leaned back. "Just one question. Why is it that you want to be an officer, and not a specialist instead, like us? How come you don't want to be a sergeant?"

"No, sir, it's not that at all. I'm a competitive person by nature, and I try to be the best. I believe OCS is the highest honor that can be bestowed on a diver, and so my natural reaction is to aspire to OCS. If I was selected to be a specialist or a sergeant, however, I would still be tremendously honored."

Warrant Foo, Warrant Jaya and Master Raj all smiled. Warrant Foo said, "Okay, Max, I'm convinced. You have me. Now, my advice to you is to get them." He gestured towards the rest of the class. "Get them to support you the way they do Flynch. The peer appraisal is coming up next week, and your score will be very important. Anything from you, Raj?"

Master Raj shook his head. "No, just keep doing what you're doing, Max. You're doing a good job."

"Thank you, sir." I picked up my rifle and walked away. Those were probably the most stressful 10 minutes I've had since enlisting.

THE IMPORTANCE OF AWARENESS

The rain lasted until nightfall. We took a trek through the jungle, practicing our contact drills in the dark. There were puddles everywhere, so most got soaked. A few of us, including me, accidentally lay down on some vines covered with thorns. They were a bitch to remove from our hands and uniforms.

Thomas and I drew the 4 am shift for sentry duty.

I'm not looking forward to it.

SOCK IT TO ME

Tonight, I cemented my place as worst tent mate of all time. Sometime in the night's murky middle, I dreamed

I was falling. In my dream, I flung my arms to the side to try to catch myself. The problem was that I also made the motion in real life. I whacked Thomas in the chest, waking him up again. If we have another field camp, I sense I'll be sleeping alone.

THURSDAY, APRIL 25

TEST DAY

Today, we had a final evaluation, covering all the drills we've learned. We were worried about the artillery drill. Because of yesterday's rain, our section hadn't learned it. It turned out to be the simplest, however. When a shell is inbound, everyone hits the dirt. Once it explodes, everyone runs the other way. That's the entire drill.

We spent the morning in review. For our final practice, we were issued blank rounds to shoot, which was pretty fun. What was less fun was the half hour we spent picking up our blank casings. While searching for the spent shells, I came across a dead scorpion. It was over 10 cm long, with a thick, bulbous stinger. I'm glad it wasn't alive.

Our test was in the jungle near our campsite. So far, we've only practiced the drills in open spaces, so the new terrain is a challenge. Master Martin briefed us. There was an enemy concentration somewhere in the forest. Our job was to eliminate it. Just before we started, we passed a distraught Platoon 2 Section 1, who told us they had failed twice. They were waiting for their third attempt.

Shit, they failed twice? Maybe this won't be easy as I thought.

We started down a narrow pathway into the jungle. Shiva nudged me to the front and told me to lead our section. I gulped and agreed. I signaled for a staggered formation, with half the section on the left side of the path and half on the right. We moved quietly, scanning our surroundings. From the right — *bang!* Blank rounds exploded through the jungle. We dropped and crawled into a firing line while providing suppressing fire.

Directly to my right was our section IC, Jared. He's socially awkward, so Instructor Shane appointed Jared IC to bring him closer to us. He would be issuing all the commands.

We lay prone in our firing line, simulating fire by yelling "bang." The next step was to leapfrog towards the enemy. I looked at Jared. Instead of checking to see if we were in position to begin, he was just looking straight ahead, shouting, "Bang, bang, bang!"

I screamed, "Jared, leapfrog!" He stared at me blankly for a second, then finally woke up. He shouted, "Group one, suppressing fire, group two, ready?"

"Ready!" echoed group two.

"Go!" yelled Jared, and we were moving.

Our targets rose into sight before us: Instructor Zhong and Master Dennis were seated about 50 m ahead. Jared shouted, "Section 4, prepare to charge. Ready?"

"Ready!"

"Charge!"

We stood up and converged on Master Dennis and

Zhong. The right side of the firing line began curling around to get clear shots, and that was our mistake.

Master Dennis bellowed, "Stop! Gather!" The test was over. We crowded around him, anxious to hear the verdict.

"You all did a pretty good job," he said. "It was a bit slow at the beginning, but your firing line was solid and you advanced well. The problem was this, at the end. The right side of your firing line was coming this way to shoot at the enemy, but look who's directly behind the enemy! You were nearly shooting at your own men! What, you want to kill your own guys?

"You need to make sure you have a clear line of sight. What you should do is have the left side of your firing line take care of this enemy. Instead of curling around, the right side should continue forward. That way they can attack any enemies deeper in the forest.

"Overall, good job. You passed. Now get lost."

Elated, we sprinted out to the main road and congratulated each other. We swaggered past Platoon 2 Section 1. They were shocked that we'd passed on our first try.

When we sat down in the training shed, I felt a strange tingling across my stomach. It was prickly and painful, as if splinters were embedded in my skin. Thinking they might be thorns from a plant, I scratched my nails over the red area, hoping to claw them out. That didn't help.

At the beginning of the week, Warrant Foo had

warned that we would all get heat rash at some point. There was no avoiding it. I assumed heat rash was an itch, though; not something painful. I asked Jan Jan, one of my section mates, if he'd felt this too. He said, "Yeah, fricking heat rash! It's like glass shards in your skin. It sucks, man."

⋇▬ ⎯⎯ ▬⋇

Eventually, every section passed the test. After lunch, Warrant Foo announced, "Gentlemen, listen up. Now that you've all finished the test, I can tell you that it was the final evolution. As of now, your field camp is officially over. We will be returning to NDU tomorrow morning." We clapped and cheered.

We relaxed in the training shed. From behind, Aneirin prodded me. "Hey, Master Dennis just talked to me, and he wants to see you."

I grabbed my rifle and hurried over to the instructors' table. Master Dennis waved for me to sit. He said, "Max, so far, you've done a great job. Not just in field camp but in general. Of all the recruits, I would say you're in the top three. I think you have a chance to be Best Trainee. The way you carry yourself, how you speak, your manner is very courteous. It's all impressive.

"My advice for you is to get *them*"— he pointed at the class — "to like you. The peer appraisals are coming up, and that's a big thing. We don't know how you are with them when we're gone, so we take the peer appraisals very seriously. Maybe you act great with us, but when we're gone, you're a cock to them. Make sure you're nice to everyone. Okay, keep up the good work."

Astonished, I thanked him, remembered to grab my rifle, and returned to my seat.

The rest of the afternoon was leisurely. We filled in our shellscrapes and planned our skits.

39'S GOT TALENT

Each section prepared a funny skit. It's a field camp tradition to hold performances on the last night.

We finished dinner at the campsite and trooped back to the training shed. Our instructors were looking forward to the show. First up: P1S3, Aneirin's section. They did a solid skit about hunting down a drug lord.

Next was P1S1, who did a disturbing all-male striptease. P1S2 followed with a story about a maid defeating foreign invaders. Next was P1S4, my section. We tossed up a sketch called "Eight Ways to Lose Your Rifle." It drew a few laughs. Elston's section followed with a show about NDU recruits trying to pick up girls in a club. Then, the last act finished the night spectacularly.

Jeffrey Hoe has an uncanny ability to mimic people. Not only voice, but mannerisms, movements; everything. His section's skit was essentially The Jeffrey Hoe Show. Jeffrey did an absolutely uproarious impression of Instructor Choi teaching us about our rifles. He had us laughing so hard we had a core workout. We were jumping up and down. I looked over to see the instructors gasping for air, Choi included. When Jeffrey yelled Choi's signature line — "If you screw with your weapon, you screw with me!" — we all died.

It was a great way to end field camp. Even better, we had already filled in our shellscrapes, so we didn't have to do sentry duty. Hooyah.

FRIDAY, APRIL 26

I awoke in the murk between darkness and dawn. A grumble of thunder tumbled over us. Two minutes later, it began to sprinkle. *Oh well, maybe it'll pass quickly*, I thought. I closed my eyes. Thirty seconds later, the sky unleashed a freezing rain, a hail of watery bullets. We heard a shrill, extended whistle blast: the signal to muster on the road.

Thomas and I threw on our gear. We jogged to the training shed as the frigid rain fell. It was the coldest I've ever felt. We arrived at the shed, soaked and trembling, and sat there for an hour and a half until the rain stopped. I lay down on the floor next to the benches. Using my LBV (load-bearing vest) as a pillow and tucking rifle between my legs, I slept.

I woke up to find the downpour was actually a blessing: my heat rash was completely gone. We dismantled our bashas, packed our field packs, loaded up the tonners and rolled out of the forest in time to eat lunch at NDU.

The afternoon went to unpacking the tonners and cleaning our rifles. It was 1700 when we finished. Only then did I remember that it was Friday — we were going home. Warrant Foo told us to get out of camp by 1800, so we rushed through dinner, skipped showering, and packed as quickly as we could.

Elston's mom picked him up and he offered me a ride. I got home, took a lovely shower, and crashed.

WEEK
SIX

SUNDAY, APRIL 28

After a sweet, sweet weekend that was all too brief, we walked back through the good old gates of Sembawang Camp. I was thankful field camp was over, but now it was back to the regular grind.

Instructor Thaddeus was the duty instructor for the night. Twice, he called us to muster down *after* lights-out. The first time, he caught someone outside his bunk. I was actually glad he did—I was in the shower when lights-out was called, so I was stuck. I was trying to figure out how to sneak back to my cabin when Thaddeus spotted someone else.

After a few sets of push-ups in our pajamas, Scott gave us one minute to return to our bunks. We made the timing, but he then snuck into a cabin and caught someone using his cellphone, so we had to muster once more.

MONDAY, APRIL 29

SOC IT TO ME

Nothing beats a timed obstacle course on a Monday morning. Racing against Benjamin and Thomas, I sprung out to an early lead and held on, flying through the obstacles and giving it all on the final sprint. I was trying as hard as I could, partially to win a bet with my cabinmates, but also to beat Aneirin's time. Flynn and I push each other to compete on each evolution, for which I'm glad.

We mustered at the training shed to hear our results. To pass, we needed to clock 4:30 or faster. Last time, Instructor Khong challenged me to hit 3:30. Thanaraj told us a friend of his in infantry had clocked in at 3:03 and was the fastest in his company. I was surprised when the first few guys all scored below four minutes, including a stellar 3:16 from Aaron Ang. Aaron is 21 but, like me, he has a baby face, so he looks younger.

I mentally took note of the fastest times. By the time the instructors had read the times of two-thirds of the batch, Aneirin was fastest at 3:07. I was nervous. I was sure I passed, but did I beat Instructor Khong's challenge? Did I beat Flynn?

After forever, Master Dennis reached my name. He called out, "Max, 2:50."

The class clapped and cheered. I was happy that I'd done well, and touched that the class supported me. Some guys tried to start a chant of "OCS, OCS." I hope I deserve their confidence.

I knew I wouldn't be the fastest, though. That had to be Ivan, the national runner. Indeed, Ivan clocked 2:42. However, Master Martin revealed that he'd spotted Ivan skipping one step on the stepping stone obstacle, so he was disqualified. That left me with the fastest official time. Hooyah, bitches.

We had a canteen break, spending 20 minutes in the Nee Soon cafeteria. I snagged myself a chocolate milk and a plate of noodles. Mmm, noodles.

APPRAISE THE LORD

We sat in the NDU auditorium, a piece of paper before each of us. Next to the list of names, there were two blank columns. The first was labeled "Command Potential." The other was "Resilience."

We had begun the class peer appraisal. We were instructed to grade each of our batch boys on a scale of 1–7, seven being the highest. I took my time, going through everyone carefully. In the end, the only guys who I scored as "7" in both categories were Aneirin, Raymond Hoe and Jay Yang.

Instructor Phillip was our DI. He gave us 30 minutes to relax before lights-out. I did pull-ups, squats, and core exercises with Aneirin and Brendan Conceicao. Brendan and I have become close recently. Brendan is one-eighth Portuguese. He looks vaguely Filipino. For some reason, this bothers him. Naturally, as a result, we make fun of him a lot.

O STANDARDIZATION, WHERE ART THOU

We have a route march tomorrow. After dinner, we prepared our field packs. Our instructions were to muster for an inspection in 30 minutes, wearing PT attire.

I messed up. I was packing my field pack in my cabin when we heard the ICs call from below, "39, muster down now!"

I hustled to find the right clothes. When we wear PT attire, we're supposed to wear green socks, but I couldn't find a pair. I asked, "Do we need to wear the green socks?"

My cabinmates were ready to go. None of them were wearing green socks.

Benjamin said, "I don't think it matters."

"Me neither," Noel added.

"Yeah, we'll all be fine," Thomas agreed. I put on some ankle-length black socks and we rushed downstairs.

Instructor Phillip seemed unhappy. We mustered in our PT formation, in 10 rows, and laid out our field pack items. Leonard checked everyone, frequently dropping people for forgetting items or for arranging them improperly.

Phillip finished his check. He told the ICs to put us in *senang-diri* position: feet shoulder-width apart, hands clasped behind our backs. This is our relaxed position. It's typically what we assume before being dismissed. I was relieved that the inspection was over. I was still worried about my socks.

"Those not wearing SAF socks, come up to the front now," Phillip boomed. "I saw a few of you. Come up!"

Oh, shit.

I thought about risking staying put. *Maybe Phillip didn't notice me. If I don't move, I might escape.* If I was caught trying to hide, though, the punishment would be far worse.

I jogged to the front of the class. I was joined by Benjamin, Noel, Thomas, and no one else. My cabinmates and I were the only ones.

Phillip said, "Look at these guys. I try to be nice, and let you guys muster in your PT kit, instead of number four. But I guess you guys don't like the PT kit. Is that right?"

"No, instructor!" the class yelled back.

"Because of these four guys, you have seven minutes to change into long fours and muster back down. Time starts now."

Everyone ran for the staircase leading up to our bunks. As we headed up, Siu Hoi muttered, "Good job,

Max." I turned and smiled, assuming he was joking, but he was serious.

Phillip had us repeatedly pack and unpack our field packs, trying to meet a timing — the equivalent of a change parade. He then dropped us for three sets of 35 push-ups.

The punishment wasn't severe. I was expecting worse. The most damaging part of it, I felt, would be the change in the way people viewed me. The class was punished for my cabin's mistake. I need to avoid doing this again.

Instructor Phillip released us with a half hour until lights-out. I did pull-ups, squats and some core exercises with Aneirin, burning off some of the frustration from my mistake. As we did pull-ups, Phillip came up to me and said, "Max, I'm disappointed in you. Why couldn't you just wear socks?"

I had no answer and offered none. I just want to stop disappointing people.

TUESDAY, APRIL 30

Today, Batch 38 officially graduated from the Combat Diver Course, or CDC. Today, they officially became naval divers. Their ceremony took place on the NDU football field at twilight. There were four large drums in the middle of the field. To the beat of the drums, the graduating class marched onto the field in their boat crews, hefting inflatable rubber boats overhead.

The 38th Batch marched in. It was completely dark by now. At the far end of the field, trip flares ignited. Two fiery columns of light rocketed towards the sky, prompting gasps from the crowd.

The drums began again. After demonstrating boat exercises, the 38th Batch mustered neatly before the grandstand. The ceremonies began, concluding with the announcement of 38's Best Trainee and Best PT. Both were spectacularly honored.

The same awards, Best Trainee and Best PT, will be presented to 39 at the end of our BMT.

LTA Jackson announced, "Ladies and gentlemen, Batch 38 of NDU!" The newly-christened naval divers threw their jockey caps into the air and celebrated.

It was a grand graduation. I hope to be part of Batch 39's.

WEDNESDAY, MAY 1

I love public holidays. We booked out last night after 38's graduation.

THURSDAY, MAY 2

We saddled up and our tonners rolled into Nee Soon. It was SOC time.

Only a few guys had yet to pass, but everyone ran the course anyway. Master Dennis decreed, "Everyone will do the SOC twice."

I ran hard both times, and was feeling the burn. Following a 10-minute rest, Master Martin delivered a not-so-pleasant surprise—everyone would be going a third time. I thought we were done, and had mentally checked out. *I already ran hard twice,* I thought. *This time, I'll just take it easy.*

Then, Master Dennis said, "Ivan, Flynn, Max. Up here, now." We jogged to the front. Master Dennis continued, "I am bored. You three are to race for my enjoyment. First place will receive a good deal. Second place will get something bad. Third place will get something worse. Is this clear?"

The class clapped and cheered. I just laughed.

Ivan was the easy favorite, but I didn't want to make it easy for him. I knew he would definitely outrun me on the final 300-m stretch, so my only shot was to destroy him on the obstacles, build a lead, and try to hold him off.

Ivan and I quickly pulled ahead of Aneirin. Sprinting, I raced ahead of Ivan on the low rope, but he overtook me on the balance beam. We began the final stretch at the same time. I was toast. Ivan bested me by a solid 10 seconds and I beat Aneirin by about the same. The class applauded as we finished.

With a little apprehension, Aneirin and I approached Master Dennis to learn what our punishment would be. He smiled and told us to get lost.

POOL SHARKS

In the afternoon we hit the pool, attempting the 500-m sidestroke swim, one of our vetoes. With fins, we have to complete it in 12 minutes and 30 seconds. I finished with a time of 8:56, edging out Aneirin and Ivan, who both clocked in at 8:57. Elston swam 8:30. Jay Yang, a strong swimmer, coasted to an 8:15. Vincent, the best swimmer in the class, finished in 8:00 — without fins.

⇢⊨◉ ⸻ ◉⊨⇠

As I've mentioned, Jay is one of the best guys in the batch. At dinner, I jokingly stole Jay's apple. When I returned it to him, he sincerely offered it to me. The guy is too nice for words. Jay is interested in signing on. His parents met in the military. His dad is a 1st Warrant Officer, and his mom was formerly a Master Sergeant. I asked him once if his parents were proud of him being in NDU. He said, "Yes, very proud. My dad actually wanted to be a diver, but he wasn't selected, so he's very proud that I'm here."

Jay pointed out a hilarious possibility. If he's sent to OCS and becomes a commissioned officer, he'll outrank both his parents. His mom and dad will have to call him "sir." How great would that be?

FRIDAY, MAY 3

Today we celebrated the 43rd anniversary of the Republic of Singapore Navy. The celebration was at Changi Naval Base. Happy birthday, RSN.

THE CURRICULUM

I've now spent enough time in NDU to broadly understand the life of a diver.

As with all combat-fit National Servicemen, NDU recruits initially undergo nine weeks of Basic Military Training, or BMT. The focus is on fitness, jungle warfare, and weapon familiarity. NDU's BMT is reportedly more difficult than the standard infantry program. Our PT and runs are tougher, and we focus on swimming as well.

Next is the 16-week Combat Diver Course, or CDC. Our physical training will intensify and we'll learn to dive with both commercial and stealth gear. We'll also undergo operational training. At some point during CDC comes Hell Week. The name and concept both borrowed from the US Navy SEALs, Hell Week is five days of nearly continuous physical exertion and extreme sleep deprivation. During Hell Week, divers receive less than five hours of sleep — in total.

Those who survive Hell Week and complete CDC will graduate as naval divers. The class will then be divided. Most will go to the Underwater Demolition Group (UDG), with the rest assigned to the Clearance Diving Group (CDG). A selected few will go to OCS.

UDG's emphasis is eponymous. Divers carry on with advanced combat diving and underwater demolition. About 70% of divers are typically assigned to UDG.

CDG focuses on RSN diving operations, search and salvage missions, humanitarian assistance, and disaster relief.

The road ahead is long and steep.

WEEK
SEVEN

SUNDAY, MAY 5

I spent Friday relaxing and Saturday night having drinks with a few batch boys.

I weighed myself before booking in tonight. After enlisting at 78 kg, I now weigh 72. Shit. I've definitely lost some body fat, but I was already skinny to begin with. Before beginning NS, I consistently worked out, but I've now completely stopped. I don't want to keep losing muscle mass, though. I'll bring in extra protein powder this week.

After booking in, we had a stand-by bed. For an unsatisfactory performance, we were dropped for 80 push-ups. Following the water parade, I did some pull-ups and dips before going to sleep.

MONDAY, MAY 6

We spent the morning swimming at Nee Soon, doing freestyle sprints under Master Martin. We divided into four rows. One row sprinted two laps of freestyle while the rest did varying isometric exercises, such as holding in the push-up and squat positions. After the swimmers finished their laps, we swapped with them.

In the afternoon, Master Dennis conducted routine interviews, checking on every recruit, making sure we were all right. At 1600, back in NDU, we went for a 4-km run. The run ended at the Grinder, where we each did one set of pull-ups to failure. I surprised myself by managing 18, my personal best.

The water parade was held at 2030, the earliest ever. Lights-out was at 2200.

Water parades are meant to ensure we're hydrated, but forcing down a full liter of water can be difficult. Thanaraj can verify that. While trying to finish his bottle, he threw up.

MELLOWDRAMA

After water parade, Aneirin asked me, "You know what Master Dennis said to me during my interview?"

"No, what?"

"He said, 'Max has mellowed out recently.' I guess he means you haven't been as active."

"Wait, really?"

I was completely surprised, but I thought hard about it and realized that I have become quieter over the past couple of weeks. It wasn't a deliberate decision; I've just gradually become less involved. Aside from being IC, there aren't many leadership opportunities. Aneirin and Ivan, however, have found ways to stay active. They urge 39 to follow instructions, and lead songs or call out the cadence when we march. I don't do as much. Strangely, I think the reason for my recent complacence is my early success. At first, I felt pressure to prove myself. I needed to stand out. When 39 started thinking I'm headed for OCS, though, the pressure lessened and I relaxed.

Back in our cabin, I asked Benjamin, "Hey man, do you think I've been quieter lately?"

He instantly exclaimed, "Yeah! When you were IC, I honestly thought you did a better job than Flynn. But since then, you've disappeared."

I wonder if it would be nicer to sink into anonymity and stay there, in quiet comfort. It would be easier. All of me wants to be a diver, but part of me would prefer to be one of the crowd instead of the one at the front. What I need to remember is that I can be more than that. I *want* to be more than that. I'll never know what I can do unless I try.

I want to succeed. I want to lead. I want to be Best Trainee. Most of all, I want to believe in myself.

TUESDAY, MAY 7

A slow 4-km run filled the morning. My right Achilles tendon is sore, and I'm not sure why. Asking around, I learned that Ivan has the same thing. It's a dull pain near the spot the tendon joins the heel bone. I was able to finish the run, but it's a little painful.

We then had a surprise stand-by bed. We didn't have a chance to tidy up at all. Predictably, we got absolutely annihilated. Our cabins are supposed to be kept neat at all times, but like nearly everyone, Benjamin, Thomas, Noel and I only clean up before inspections. Our beds were unmade, our cabinets were messy, and our duffel bags were strewn across the floor.

Master Dennis strode into our room. I was nervous as hell. We knew our cabin was a mess. We heard several thumps and whoomps, and then Master Dennis emerged.

"Ten minutes. Tidy up. Move!"

Master Dennis had tornadoed our cabin. Our bags were emptied, our cabinet contents scattered. Noel's mattress was on the floor. Frantically, we tried to sort out the mess.

After 10 minutes, we mustered downstairs to pay the price. In total, we did 70 push-ups, 30 leg raises, 30 flutter kicks and 10 bodybuilders.

When it was done, we were in the *diam* position, ready to move off for lunch. The *diam* position is a strict at-attention. We aren't allowed to move whatsoever. Instructor Shane was about to dismiss us for lunch when someone moved.

"Drop!" screamed Shane. We dropped. As we held the position, he yelled, "*Diam* can move?"

"No, instructor!" we chorused back.

"*Diam* can move?"

"No, instructor!"

"*Diam* can move?"

"No, instructor!"

"*Diam* can move?"

That went on for nearly a minute. It actually became funny. I was doing my best not to laugh as I kept shouting, "No, instructor!"

⊷═ ── ═⊷

In the afternoon, we headed to Nee Soon for another 500-m sidestroke swim. We have an IPPT coming up in two days, so Ivan suggested that we take it easy on the swim. I agreed, coasting to a time of 9:35.

On the tonner journey back to NDU, it was raining. I wanted to sleep but wound up sitting underneath a hole in the ceiling tarp. Rainwater dripped on me non-stop. It was like some Chinese water torture bullshit.

Dinner was at 1830, with another stand-by bed at

2000. We didn't know who the inspecting instructor would be. When Instructor Russell arrived, we smiled a collective smile. Russell is the most easygoing instructor we have. He stayed true to his reputation by not checking any of our rooms. We just had our water parade and were finished for the night.

SPEAKING OF REPUTATIONS...

Sometime in the third or fourth week, I didn't shower for two straight nights. I'm not normally that disgusting; I was just really tired. Yet since then, I've been known as a non-showerer. Every night, Thomas grins and asks me, "So, Max, gonna shower tonight, or do what you normally do?" I also hear, "Hey, remember to wear socks this time?" quite often.

HELLO, LADIES

I found out tonight that Jay Yang's email address begins with "icedaddy." When he told me, I laughed for about 10 minutes. He protested, "No, no, it's because I'm *cool.* Like ice. Do you get it?"

I assured him that I got it. He shook his head and said, "Okay, you just need a cool nickname, too. How about Papa Dragon?"

Well, I couldn't say no to that. Watch out, ladies, when Ice Daddy and Papa Dragon are on the prowl.

WEDNESDAY, MAY 8

The plan was to leave for the Nee Soon pool at 0645, but our tonners were delayed. The instructors let us relax until 0930. They even granted us a canteen break. I downed a bowl of mee soto and five fried eggs. Delicious.

The instructors want us to do well on the IPPT tomorrow, so today was smooth. A lot of guys are nervous, including Aneirin and Raymond. The *ang moh* is still nervous about his jump, and Raymond's afraid he won't make 9:14.

Ivan's been training hard with Brendan, who ran 9:08 last time. Ivan predicts Brendan will run 8:45 or faster. I'm aiming for 8:30.

MANHUNT

Aneirin, Aaron and I were chatting over dinner when Aaron said, "Max, I showed my girlfriend pictures of you and Aneirin, because she was curious about you white guys. Max, she calls you the good-looking *ang moh*."

"Take that, dickweed," I yelled at Aneirin. Aaron and I high-fived.

"Please," Aneirin said. "Praveen's girl said I'm the good-looking *ang moh*."

"Say what?" I said.

We found Praveen, who confirmed the unthinkable: his girlfriend thinks Aneirin is better-looking than I am. I said, "Okay then, let's make a deal. We'll ask the batch's girlfriends who they think is more attractive. Winner gets a beer."

"Hey man, if you wanna buy me a free beer, I won't say no," Aneirin replied.

That British bastard is going down.

GENERAL MISCHIEF, SIR

While Benjamin was showering before dinner, Harry, one of our section mates, hid inside Benjamin's cabinet. Just then, Noel walked into the room. I said frantically, "Noel, someone stole everything in Benjamin's cabinet! Look!"

"Wait, shit, really?" Noel exclaimed. He opened the cabinet and Harry leapt out, screaming like a maniac.

I swear Noel nearly died. When I was done laughing, I had to physically restrain Noel from locking Harry back inside.

Later on, I noticed that one of the plasters on my feet had congealed into a sticky, grimy mess from swimming. When I rolled it up, it was the size of a marble, and looked like a giant booger. Keeping it on my finger, I strolled into Elston's cabin. The first guy I saw was Sean Teh.

I said, "Hey, Sean, do you want to see the most disgusting thing of all time?"

"Yeah, sure! What is it?"

I held up my finger and said, "Isn't this just a *giant* booger?"

He jumped about a foot backwards and shouted, "Ahh, Jesus! What the hell, man?"

Elston and Thanaraj ran over and I showed them, too. Thanaraj was disgusted. Only Elston laughed, shook his head and said, "No way, man, no way that's a booger."

"Yes it is!" I insisted. With my finger outstretched before me, I ran towards Sean. He shrieked and ran away, dashing out of his room. "No, come on, seriously, STOP!" he shouted as I pressed it on his face.

When I finished laughing, I said, "Nah, don't worry man, it's not a booger. It's just a used plaster."

"Awh, shit, that's even worse!" Sean yelled. He shook his head. "Not funny, man," he said.

THURSDAY, MAY 9

THE I-DOUBLE-P-T

The big day. The last IPPT of BMT. "That SBJ," Aneirin kept saying. "I'm gonna be so sad if I don't get 243."

I decided to offer some advice. "Just jump," I said. "Like, just jump 243. Honestly, just do it."

"Yeah, because it's that easy."

"It is," I said. Aneirin punched me.

The static stations of the IPPT went smoothly. I stopped after the requisite 12 pull-ups, did 45 sit-ups, ran a 9.2-second shuttle run, and jumped 270 cm. All that remained was the run. As before, we ran in platoon-level. Fortunately, Ivan is in Platoon 2, so I don't need to race him head-to-head.

TWO-POINT-FOUR

"Go!" the instructor called.

As usual, nearly everyone started too fast: after sprinting the first half-lap, they quickly slowed down. I began around eighth place, trying to maintain a consistent pace. For the first half-lap, I was in the middle of the group, nowhere near the lead.

By the end of the first lap, Aneirin and Aaron were side-by-side at the front, a meter ahead of me.

Aaron only ran 9:19 last time, so I kept waiting for him to fall back, but he stayed with Aneirin.

A few times, I moved to overtake them, but they regained their small lead each time. We remained close.

At the start of the final lap, I pulled into first place. Aaron and Aneirin soon pushed ahead, carrying a small lead.

I was exhausted, but I wanted to win. I planned to kick early and hold on.

With 150 meters to go, I rose into a full sprint, darting ahead of them both. *I've got this,* I thought. *I've actually got this.*

At 50 meters from the finish, I heard footsteps, then two streaks of navy blue flashed past me. Aaron and Aneirin blazed towards the finish. I tried to catch them, but they straight out-sprinted me.

I finished third. Aaron came in second. Aneirin triumphed, retaining his title as fastest in our platoon.

GOOD TIMES

Aneirin clocked in at 8:11, Aaron at 8:13, and I finished at 8:16. Raymond came in fourth at 8:32. Though I'd finished third, I was pleased. 8:16 was a sweet-ass time! It was better than I'd hoped.

Still, I wish I could've beaten at least one of those narcs. Aaron somehow improved by over a minute. Unreal.

BAD TIMES

After Platoon 2 finished their run — Ivan smoked everyone, running 7:55 — we mustered on the basketball court. All of our instructors were lined up before us. This was unusual, but I didn't think anything of it.

They gave us two minutes to change into long four. We're taking our BMT class photo today, so we thought we were dressing up for that. When we'd finished changing, though, Instructor Zhong said, "Take off your garters. Remove everything from your pockets."

Huh? We've never done that before.

Master Dennis stepped forward. He didn't look happy. "39, you all should be ashamed," he barked. "Your discipline is terrible. All of us have noticed this over the past few days, but we've kept quiet.

"You finish the IPPT, and you can't even walk over here and muster quietly and orderly. You can't even stand properly right now. All of you, DROP!"

We hit the floor. Master Dennis snapped, "You will learn to behave. Everyone to the loading bay. Two minutes. Go!"

We got up and ran across the NDU compound to the loading bay. Master Dennis pointed to the slipway, an asphalt area bordering a small beach. Beyond the short stretch of sand, the sea glimmered.

"Muster in section levels. Go!"

We scrambled to organize ourselves. Evidently, we didn't muster fast enough. Master Dennis said, "You don't even want to muster quickly. Okay, we'll do it our way then. DROP!"

Warrant Foo grabbed a nearby hose and soaked us. Master Dennis screamed, "You think this is a game? You think NDU is a joke? DOWN!"

"One!" we echoed.

"DOWN!"

"Two!"

We reached 20. Master Dennis bellowed, "On your backs!" We flipped over. Not everyone was facing the same direction.

"You can't even lie down properly!" he shouted. "On your bellies!"

We flopped onto our stomachs.

"On your backs!"

We rolled over.

"On your bellies!"

We rolled again.

"On your feet!"

We sprang up.

"Push-up position! Follow the whistle!" Master Martin put an orange whistle to his mouth. Each time he blew it, we did a push-up.

Several eternities later, we proceeded to flutter kicks, again following the whistle.

We continued, alternating between push-ups, flutter kicks and leg raises, interspersed with switching between our backs, bellies and feet. There was no silence — throughout, the instructors constantly exploded at us.

"You call that a push-up?"

"Raise your legs higher!"

"What the hell are you doing?"

"You sure you want to be a diver?"

After some time, Master Dennis hollered, "You still can't work together properly! You will take turns as sections getting wet and sandy! Section 1, move!"

We'd heard about getting wet and sandy. Section 1 stood up and ran to the sea. The rest of us continued with the exercises, fighting through the flame searing our muscles. I was actually looking forward to getting wet and sandy — it would be a brief break from the pumping.

Finally, I heard, "Section 4, wet and sandy! Go!"

"Come on, guys," I shouted, sprinting for the sand. We dove into murky green and crawled out onto freckled yellow.

"Get sandy! Cover yourselves! Roll over! Get your face! Get your faces sandy!"

We rolled around in the sand, completely coating our uniforms. I cupped sand in my hands and threw it up into my face. Instructor Wayne yelled, "Max, you call that wet and sandy? I can still see your face. Get your face in there!" After a second's hesitation, I plunged my face into the beach. Our section ran back to the slipway, soaked with seawater and caked with sand. The grains dug into my skin with every movement. With every push-up, every flutter kick, I felt them grinding into my body.

When the entire batch was wet and sandy, Master Dennis commanded us to our feet. We mustered in threes. We took a head count. All present.

Are we finished? I didn't dare to guess.

"Forty-five seconds. Basketball court. Muster. Time starts now."

Forty-five seconds? Everyone bolted off. The IC shouted for us to stay in threes, and though a few of us tried to relay the command, nobody listened.

We ran haphazardly to the court, and the first guys to arrive were immediately dropped. When I got there, I turned to see if the class was close behind. Instructor Wayne yelled, "Max, what the hell are you doing? Drop!"

We did more of the same: push-ups, flutter kicks, leg raises, switching between our bellies and backs and feet. Master Martin's orange whistle shrieked all the while.

A bodybuilder named Jack Soh was directly in front of me. When we rolled from our backs to our bellies, he accidentally kicked me in the right eye, knocking sand beneath my eyelid. I blinked and blinked but couldn't get the sand out. Opening my right eye was agonizing. I squeezed it shut, involuntary tears running a river through the sandbed of my cheek.

The basketball court is adjacent to the football field. Master Dennis shouted, "Everyone, hit the fence and muster back. One minute!"

The far end of the football field is encased by a silver chain-link fence, over 100 meters away. To "hit the fence" is to run, touch the fence, and return.

We took off across the grassy field. Some guys were really struggling. I was tired but could push on. The worst part was the sand in my eye.

We didn't make the timing, and had to hit the fence again. This time, they faulted us for not running with a buddy, so we hit the fence a third time, bear crawling back. To bear crawl, we drop into the push-up position, then move forward on all fours.

I needed a buddy. Scanning the field, I grabbed a funny guy named Ze Ken and asked, "Buddies?" He nodded. As we bear crawled, he began to falter. I said, "Come on, man, I don't know about you, but I really don't want to hit the fence again." The two of us hustled back to the basketball courts, sneaking in just before the timing.

We mustered, reported strength, and then it happened—we were done. We walked to the grandstand for a debrief.

"So, gentlemen, how was the session?" asked Warrant Jaya.

We gave the only response we could. "Good, sir!"

"Gentlemen, this little session was actually planned. We wanted to do it earlier, but because of the IPPT today, we decided to postpone it. I hope it wasn't too tiring, because compared to CDC, this is only about 20% of what you'll get."

We shook our heads. *Twenty percent?* He had to be exaggerating. That felt like death.

Master Dennis said, "Gentlemen, what Warrant Jaya is saying is true. That was nothing. It was only half an hour."

Warrant Jaya continued, "Correct. Today's session was actually scheduled to be one hour, but since you all did so well in the IPPT, we decided to shorten it. Believe us, this was nothing. If you want to make it as a diver, you will endure far worse."

THE UNEXPECTED VIRTUE OF PAIN

The pumping was effective. Suffering together brought us closer as a batch. Afterwards, we were marching in step, singing loudly, and, for the first time, taking pride in our behavior.

We collected some diving equipment before lunch, drawing masks, booties, and fins. Over our meal, Jonas Koh was having fun with Aneirin. At 21, Jonas is one of the older recruits. He's going to sign on, and aspires to be one of the few naval divers sent to train with the US Navy SEALs.

Jonas, Aneirin and I were eating together when Jonas looked over Aneirin's shoulder, straightened up and hissed, "Master Sergeant, Master Sergeant." When Aneirin turned around to see, Jonas hit him in the balls. There was no Master Sergeant.

CHECK TWICE, WIPE ONCE

We did area cleaning in the afternoon. I passed Zeng Hou in the hallway. He looked pained. I asked him what was wrong. He groaned, "I just made the worst mistake of my life. I took a shit, but my asshole was still sandy from the whacking. When I wiped, it was like wiping with sandpaper."

A stand-by bed followed. It was essentially a push-up session. Five instructors were present and they found excuses to drop every cabin for 50.

DON'T BE THE LAST ONE OUT

Before dinner, as we were mustering downstairs, Michael Li accidentally locked his cabinmate inside their room. Our doors lock from the outside. As Michael left his room, he turned off the lights and locked it. Only when we were mustered downstairs did we hear shouting and frantic banging coming from one of the cabins. We were all in hysterics as Michael ran up to let his buddy out.

SPEEDING UP

The IPPT results were posted on the board at Block 29. They were incredible. On our first IPPT, six guys got Diver's Gold. Now, there were 32. The number of failures dropped from 30 to seven.

There was bad news, however. Jonas has uprooted me from my throne as best in SBJ, as he jumped 275 cm to my 270. What a jerk.

After dinner, the instructors let us relax for the night in anticipation of our 12-k route march tomorrow morning. Our lights-out time was 2055, the earliest ever.

FRIDAY, MAY 10

"One minute. Everything on, and muster in threes!"

I checked my watch. Zero-eight-hundred.

Hefting my field pack onto my shoulders over my LBV, I reluctantly fell in. I fastened my helmet and adjusted my rifle strap. I dreaded the upcoming 12 kilometers like I used to dread wetting the bed at sleepovers.

You can book out after this, I reminded myself. *Just get through this and it's home sweet weekend.*

We began, plodding through our usual route around Sembawang Camp and NDU. We did what we could to battle the boredom and fatigue, singing and chatting. The pace was painfully slow, as we trudged at a rate of 15 minutes per kilometer. The worst burn was in the trapezius and upper back, the load-bearing muscles, from carrying the field pack and rifle.

We had a short break every 4 km. I divided the march into 4-k chunks. *Just get through 4 k at a time. Just make it to the next break.*

When we finished, it was nearly noon. I can't imagine marching twice this distance at the end of BMT. Those 24 kilometers are going to suck.

We returned our rifles and magazines to the armory and gathered on the grandstand before Warrant Foo. He asked, "Who here has Diver's Gold or BMT Gold?"

Around 40 of us stood up.

Warrant Foo said, "Gentlemen, you all can *pang kang* from this place after lunch. IC, where are you? I want you to have your men out of here by 1245. If they're not, you'll stay here till six, and book out with the rest of them. Understood?"

We rushed through lunch and ran back to our bunks to pack our bags. The rest fetched their fins for a swim at Nee Soon.

At 1315, we made it to the Sembawang Camp gate and were officially free for the weekend. I was exhausted. Ivan, Peter Wong, Michael Li and I all live fairly close to each other. They asked if I wanted to share a cab.

"Hell yeah," I said, and hopped in.

WEEK
EIGHT

SUNDAY, MAY 12

It was Mother's Day today, so we had a family lunch. There was good food and a great bill.

On Saturday night, I played basketball with Ivan, Brendan and Jeremy Tan. In the middle of a game, Jeremy inadvertently delivered a solid knee to my right quadriceps. It's really painful to walk now.

We booked in at 2000. Instructor Wayne didn't check our field packs, saving us a solid half-hour. When I walked into our cabin, Benjamin was doing laundry. He asked if I wanted anything cleaned. I thanked him and handed over my pillowcase. He asked, "Do you know how much powder I should use?"

"Not a clue," I replied. "I've done laundry as many times as Thomas has gotten Diver's Gold." (If you're keeping track at home, that would be zero.)

Benjamin laughed. Thomas punched me.

IT'S ALL IN THE LABEL

As I've been sitting in my cabin writing down these notes, three different guys—Jack Soh, Zhi Heng, and Blake—have walked in and asked if I'm writing in a diary. No, it's not a diary. It's a *journal*, dicks. It's way more manly. I swear.

MONDAY, MAY 13

Warrant Foo led the morning run. My leg is still really sore from where Jeremy kneed me, so I asked Warrant Foo if I could sit out. He agreed. I waited at the Grinder with the other injured guys. The class only ran 2 km before arriving for statics, so I didn't feel too bad for missing the run. Master Martin led us in planks, side planks and leg raises. We then did plyometrics training for the standing broad jump. I sat this out, too.

Master Dennis then called us to Block 29 for our pre-lunch ration of pull-ups. The class is split into two pull-up groups, strong and weak. Whenever we do pull-ups, the strong group now does 16 and the weak does 12. Those who can't do 12 are assisted. When it was my turn, I did 16 and was ready to drop from the bar when Master Dennis said, "No, hold it there. You can only come down when I say so."

I laughed. He didn't. I did one more, reaching 17, and was burned out.

"Permission to come down, sir?"

"No."

I hung from the bar until everyone else was finished. My forearms blazed. Twice, I couldn't hold onto the bar any longer and slipped off. I pretended to wipe off my hands, as if sweatiness was the issue, and jumped back up.

When the entire class had finished, Master Dennis mercifully said, "Down." With relief, I dropped. He pulled me aside and said, "You need to be stronger. I need you to be a solid soldier. Whatever CDC throws at you, I need you to be able to take it."

"Yes, sir."

"Go muster with the rest."

My forearms were so exhausted I could barely hold my water bottle. I tried to clench my hand into a fist, but physically couldn't.

⋅→▣⋅ — ⋅▣←⋅

Tomorrow and the day after, we'll be at Nee Soon camp for our SAR 21 live firing. After lunch, Master Raj delivered the range brief. I'm excited to shoot a real gun.

Next came a talk on the SAF Core Values from Warrant Jaya. There are eight SAF Core Values. They are loyalty to country, leadership, professionalism, discipline, fighting spirit, care for soldiers, ethics, and safety.

After the talk, Warrant Jaya hit us with a pleasant shock. He announced, "39, you all will not have your 24-kilometer route march. You will do the 16 k, and that will be your final one."

The room exploded. Some of us were in disbelief, some were high-fiving, and some were hugging, but we were all ecstatic. *No 24 kilometers?* The 24-km route march is the traditional apex of Basic Military Training.

"Uh, sir, but why?" Eric Teo asked. Everyone began shouting for him to shut up.

Warrant Jaya smiled. "Why do you need to do it? There's no point."

We returned to cheering. "Thank you, sir!"

KRYPTONITE

It was still too early for dinner, so we waited in the auditorium. There were no instructors around. I was sitting near the front when Ramesh poked me in the ribs to get my attention. I involuntarily jerked upwards. I did so because I'm ticklish. Really ticklish. Exceptionally ticklish, in fact. A fact I'd managed to hide until Ramesh's discovery.

Ramesh said, "You're that ticklish?" He poked me again and I twitched.

"Ha, ha, very funny. You can stop now," I said.

"Hey, look at this ticklish *ang moh*!" Ramesh called.

In seconds, I was getting poked from all sides. Ramesh was behind me, Elston and Sean Teh to my right, Zhi Heng in front of me, Aneirin to my left. I had no defense for such an attack. They all thought it was hilarious. Through my own hysterical laughter, I managed, "Come on, stop it!"

"Alright, alright," someone said, and the tickling ceased.

Suddenly, I was grabbed from behind. Elston and Sean Teh pinned my arms, pulling me backwards so I lay across two chairs. The rest tickled me from the front. I was laughing so hard I couldn't breathe or speak. I rolled around, trying to break free, when a loud crack burst through the room.

Instantly, everyone froze.

"Ooooh," someone said.

The back of my chair had snapped off. It was lying on the floor behind us.

"Oh, shit," said Elston.

They let me go. I got my breath back. The class started laughing.

Aneirin helpfully said, "Hey, we should fix this before someone comes back."

"No shit, you wanker," I answered.

Sean cried, "I know!" He ran over to a different chair, one that's been broken since we enlisted. There's a sign on it that says, "PLEASE DO NOT SIT." We shifted the sign onto the newly broken chair.

"Problem solved," said Sean.

TUESDAY, MAY 14

LIVE RANGE, DAY ONE: "SHIT"

We woke up at 0430 to leave for the Nee Soon live firing range. We filed into the outdoor facility around sunrise. Twelve firing lanes were split into two chambers, six lanes a piece. There was a large training shed. It felt like field camp.

The instructors were loading magazines with live rounds as we marched in. Blank rounds are black and rubbery, but the live rounds gleamed gold in the rusty early-morning light.

We have eight practice shots before the test begins. To pass the live firing test itself, we need to hit 16 of our 32 shots. Those who fail will re-shoot.

We fired in section level. When Instructor Khong called "Detail 4," my section rose to collect our magazines. Each time we collect or return a magazine, we have to write our rank and full name and sign for it. This quickly becomes tiresome if your full name is long—say, perhaps, if it's Maximillian Alisdair West.

I was in Lane 9. I stepped down into the foxhole and pulled my sandbag into position. I laid the barrel of my rifle atop it. The foxhole reached my chest. Because we're standing and we have a sandbag for support, the foxhole supported position is the easiest to shoot from.

Our instructors conducted the range from a command center behind us. Tinny speakers relayed Master Raj's voice across the range.

"Firers, magazine of four rounds, load and make ready."

My hands were shaking a little as I loaded and cocked the weapon.

"Firers of four rounds, watch your front."

The command to fire.

I clicked my weapon from "safe" to "fire" and peered through the scope. A Figure 11 target, the size of a child's coffin, sprung up, one hundred meters away. If the target was hit, it would fall and then rise back up. If the target didn't move, the shot had missed.

I was nervous. I was acutely aware that in my hands was a real gun with real bullets. If I wanted to, I could kill someone. I held in my hands the power to bring death.

Peering through the scope, I took a deep breath, and squeezed the trigger slowly. A bang, a slight recoil — and the target fell. *Sweet.*

I hit three of my four shots from the foxhole. We switched to the prone unsupported position, lying on our stomachs, propped up on our elbows.

"Firers, watch your front," said Master Raj.

I clicked my weapon to "fire" and exhaled.

Four shots later, my target had only fallen once. Overall, on the practice round, I had hit four of eight shots. I didn't feel bad, though. Not yet. This was just practice.

As I climbed out of the foxhole, Warrant Foo asked me how I shot. I smiled and answered, "Not very well, sir. Only four out of eight. I'll do better next time."

He replied, "Good, because this will dictate whether you graduate at the top of your class. The live firing is worth about 30% of your overall BMT score. You better do well. Understand?"

Thirty percent of my overall BMT score? I swallowed hard. "Yes sir," I said.

After lunch, the test shoot began. There are 32 shots in total: we fire 16 in the day and 16 at night. For both the day and night shoots, our first 12 shots are from 100 meters, and the final four are from a standing position 50 meters away.

THE DAY SHOOT

It was an absolute disaster. Our three firing positions were foxhole supported, prone, and squatting/kneeling. We fired four shots per position. I shot four out of 12.

Four out of twelve? That's not even passing. I'm on track to fail this test.

I sat in the training shed, seething. My confusion swelled into a rancid mix of anger and disappointment. I sucked. And I didn't know how to get better. I mentally replayed my shots, thinking about what I'd done, but I couldn't diagnose my problem. My hands felt still, and

I thought I was squeezing the trigger slowly, so I wasn't jerking the gun. A few guys had defective rifles, causing them to consistently miss in one area. My misses weren't bunched together, though, so I don't think my rifle is at fault.

Four out of 12? What the hell?

There was about an hour's wait before we shot our final four rounds. I sat alone and simmered. I guessed it was a mental problem. I had been nervous. I knew I was too fixated on the fact that I was holding a real gun with real bullets.

I picked up my rifle and knelt outside the training shed. I stared through the scope, holding my breath, firing mental bullets. I visualized the target dropping and rising up again. *Hit. Hit. Hit. Hit.* Hitting my next four shots would lift my score to 8/16. I would be on track to pass.

We walked halfway out onto the range, fifty short meters away from the target. I felt good. I had muscle memory, and I had confidence. I was ready to hit all four shots.

Bang, bang, bang, bang—

Four pulls of the trigger later, my target had fallen only twice. Two out of four. I hurled mental obscenities at myself. My final day shoot score was 6/16.

Awful.

Pathetic.

We gathered in the training shed for lunch. Master

Raj asked those who failed to stand up. I stood. So did six others. It was embarrassing.

I sat on the end of a bench and ate alone. As I did, Master Dennis, Warrant Jaya and Master Raj called me over. I knew they were going to ask me what had happened. *Don't show your disappointment,* I thought. *Act confident.*

"Max!" Warrant Jaya barked. "You failed? What's going on?"

"Don't worry, sir. I'm just getting the bad shots out of my system," I said.

Warrant Jaya laughed. I continued, "I am a bit frustrated, but I'm sure I'll do better in the night shoot." Warrant Jaya nodded and dismissed me.

My answer satisfied the instructors, but it wasn't true. I was struggling, and the worst was not knowing why. I was shaken.

THE WAITING

In the afternoon, we had nothing to do but wait for nightfall. I swiveled between anger and dejection. I wrote earlier that a soldier is nothing without his weapon. What is a soldier who can't shoot? I did my best to hide my frustration. I only talked about how I felt to Aneirin.

Before the night shoot, the instructors tested the weapons of the "bobo shooters," those of us who'd failed.

I handed over my weapon to Instructor Choi. He finished firing and approached me. Smiling, he said,

"You're screwed up, Max. I shot 75 rounds with your weapon and only missed five or six times. You're screwed up." Everyone around me laughed. Some loudly.

We continued the wait. Master Raj came over to check on us, asking a few routine questions.

"Is everyone okay?"

"Yes, sir," we chimed back.

"Anyone need to see the medic?"

"No, sir."

"Is anybody struggling?"

Before we could answer, Rahul yelled, "Max West!" The batch laughed. It stung. I wouldn't have cared if someone I disliked had said that. But I had considered Rahul a friend.

A few minutes later, I sat at the end of a bench, visualizing hitting my targets on the night shooting. One of my platoon mates, Jason Chao, nudged me and said, "Maxi, are you sad?"

Apparently, I wasn't hiding my feelings that well. "Yeah, I'm disappointed," I said.

Jason said, "So, on the IPPT, you got third place in the 2.4. That must be annoying. Are you mad about that too?"

What the hell? I didn't know why Jason was baiting me. I said, "No, man, I'm not mad. I improved by 34 seconds, so I'm good with that."

"What was your time again?"

"8:16."

"Well, you still got third," he said.

By now, I was pissed off. I asked, "Hey, what was your time?"

Jason said, "I got 9:42."

I smiled and said, "That's cute."

Jason turned to his friend Desmond and exclaimed, "Hey, look at Max bragging about his 2.4!"

Thomas was sitting in the row ahead. Jason prodded him and said, "Hey, Thomas, you're vegetarian, right? You can run fast, but can you jump? I can jump 250."

By this point, I was furious. Jason was being a huge dick for no reason. Thomas politely replied that he hadn't jumped far enough for Diver's Gold.

"What about you, Max? How's your jump?" Jason asked.

"275," I said.

Jason again wheeled to Desmond and said, "Whoa, look at Max bragging about his SBJ!"

I wanted to fight him, but I just turned away.

THE NIGHT SHOOT

The darkness settled in. Master Raj said, "Okay, all of you, listen up. The night shoot will once again be four mags of four rounds. This time, the first round in every mag will be a tracer. The tracers are very cool. It's like something from *Star Wars*." We laughed. "No, really!" he insisted. "Like *Star Wars*."

I didn't really believe him, but when the first detail

began to fire, we gasped. The tracers were amazing. The bullets trailed streaks of furious red, blazing into the sand bank on the far side of the range. Occasionally, a bullet would flame for a few seconds after impact. Incandescent lasers streamed through the night.

Eventually, Instructor Khong called, "Detail 4, collect your magazines!"

I rose.

For the day shoot, I had dampened my frustration, trying to stay calm. As I approached the range, however, I thought, *Maybe I should do the opposite. Maybe I should focus on my anger. Feed off it. Use it as fuel.*

As I walked to the firing point, I was struck by a desire to write. Maybe I could melt into my rifle, slip inside its barrel and become the tracer itself, a molten burst, exploding into the target. I saw myself transforming into that crimson rage, singeing the air as I burned by, streaking through the heart of the target.

The first 12 night shots were from 50 meters away.

"Firers, watch your front."

From the prone position, I clicked off the safety and leveled my weapon. My target sprung up before me. I held my breath, and squeezed the trigger.

I missed my first two shots of the night.

Instantly, my fury evaporated. The false fortress I had built from anger caved and collapsed. From the rubble rose only desolation, the feeling of failure, bleak and familiar. *Screw this,* I thought.

Then I hit my last two shots.

A spark of determination.

On each of the next two rounds, I hit three of four shots. So far, I was 8-for-12. I was feeling a little better.

We collected our final magazine of four rounds and climbed into the foxholes, 100 meters away.

I hit my first three shots. There was commotion behind me, though. My firing assistant was Richard Lum. I heard him yelling about something. I couldn't make out the words through my earplugs, though, so I decided to ignore him. I missed my last shot.

I said, "Hey, Richard, I couldn't hear you. What were you saying?"

He said frantically, "Max, you were shooting at the wrong lane!"

I blinked. "Come again?"

It turned out I was shooting at the wrong target. I was in lane 9, but had fired into lane 10 — my cabinmate Benjamin's target. Consequently, Benjamin had thought he was pointing at the wrong target, and fired into lane 11. Basically, I had personally caused a huge cluster-foxtrot.

Oops.

Eleven of my 16 shots had been on target, but as my final three hits were on the wrong target, my official result was 8/16. Combined with my day shoot score, I had shot 14/32.

Failure.

As we packed up and prepared to leave Nee Soon, Master Dennis approached me and told me I will re-shoot tomorrow. I nodded. I will do well. I have to.

When we got back to camp, we were exhausted. We just wanted to sleep, but we had to return our rifles. On our way to the armory, Instructor Choi dropped us for moving too slowly. *Come on, man,* I thought. *Give us a break.*

It's 0045 right now as I write in the dark, 45 minutes after lights-out. That's a new record.

WEDNESDAY, MAY 15

LIVE RANGE, DAY TWO: HOME ON THE RANGE

In the morning, we headed back to Nee Soon.

There were only two re-shooters from Platoon 1, Harry and me. A few others had managed to scrape the passing mark, scoring 16 or 17 out of 32. Since they had passed, they weren't re-shooting. Though I had failed, I was actually glad to be shooting again. I'd been embarrassed, picked on, and laughed at, but at least I had a shot at redemption.

Thirty-two shots, to be precise.

Aneirin offered me his rifle to shoot with today. He named her Elizabeth, after the Queen. He shot 29/32 yesterday. I accepted.

As I left the training shed to prepare for the shoot, a few guys came over to wish me good luck. Rahul was one of them.

I descended into the foxhole and propped my rifle on the sandbags. I realized my hands were actually trembling. What if I failed twice? How embarrassing would that be?

As I stood in the foxhole, a revelation crystallized in my mind. Yes, fear can ruin me. But fear exists only in the mind. If I am fear's creator, I can also be its destroyer. The thought steeled me. *Fear is a product of the self. What if I simply choose not to be afraid?*

Master Raj's voice carried over us. "Firers—watch your front!"

Calm and ready, I stared down my target. *I can do this.* I centered the scope, exhaled, and squeezed the trigger.

Misfire.

I raised my hand and called, "IA, IA, IA." I performed the immediate-action drill, reloading my magazine and cocking my rifle. I glared down the scope, my crosshairs centered on the target's chest. I breathed in slowly, and held my breath. *I can do this.* I pulled the trigger.

Misfire.

"IA, IA, IA," I shouted. I performed the drill again. I stared through the scope. I exhaled. I squeezed the trigger.

Misfire.

Are you kidding me? Just what I needed—three straight misfires.

Instructor Choi shouted, "No, no, Max, stop!" He explained my mistake. After cocking the weapon, I was releasing the charging handle under control, instead of allowing it to snap back naturally. By the time I'd fixed the situation, however, everyone else in the detail had finished firing. I hadn't fired a single shot. I was now shooting completely alone. *Great, just what I needed. Literally everyone staring at me.*

Instructor Choi whispered, "Take your time."

I stared, breathed, and squeezed. I pulled the trigger four times. The target fell four times. I shot 4/4.

After my fourth hit, Instructor Choi slapped me on the back and shouted, "Yes!"

We transitioned from the foxhole into the prone position.

I shot 4/4.

Holy shit, I thought. *I'm eight-for-eight.*

The third position was kneeling. I missed my first shot. Panic surged through my chest.

I looked away and took a deep breath. I put my rifle to safe.

Brendan Conceicao was my firing assistant. I asked, "Brendan, where did I miss?"

He yelled back, "I don't know. I didn't see!"

Okay, whatever. I can do this. I looked back downrange, flicked off the safety, and hit my next three shots.

I swelled with exhilaration. *Hooyah hey, I don't suck!*

We fired our final four rounds from the standing position, 50 meters away. I hit all four, notching a total of 15/16 on the day range.

I got back to the training shed. Aneirin found me and asked, "So? How was it?"

I grinned. "Fifteen out of 16, son."

He lit up and slapped me on the back. "I knew it! Congrats, man."

I sat down, trying to hide my giant grin. I thought about how I had fired so poorly the first time. The simple explanation was that I'd changed rifles, but I didn't think that was it. After all, Instructor Choi had hit 70/75 with my weapon. I sensed there was a different reason. Yesterday, I think I might have been flinching as the trigger reached the biting point, just as the weapon fires. I think I was closing my eyes and jerking the trigger, just enough to miss. Today, though, whether it was the added experience or the added calm, I was able to keep my eyes open and stay steady. For the first time, I felt confident.

"Max, come here." Warrant Foo beckoned me over.

I walked over. "Sir?"

Warrant Foo said, "In life, there will always be external factors you can't control. When things like that happen, you can't let them get you down. Look at yesterday. Maybe you had a bad weapon. But why did you get that weapon? Exactly — you don't know. You can't control it. So when that happens, sometimes you just have to accept it and move on. Yesterday, how did you feel?"

"Very frustrated, sir."

"I know. You were dragging your face around all day!"

I laughed. Evidently, my poker face lacked conviction.

"The entire day, you with your sad face. But did it help? You need to accept things and move on. If you can do that, people will think, 'Hey, this guy's been through struggles, but he can pull himself together. He can carry on.' That's what you have to do. Clear?"

"Yes, sir!"

"It was a good experience for you to have. You did well to come back today. Yesterday was the first time you struggled, right? Yes. It was a new feeling. Next time, the feeling won't be as bad, because you've experienced it before. I believe yesterday was a better experience for you than today was. Understand? You have more peaks to hit in CDC. Keep working hard. Carry on."

I returned to my seat.

Just before lunch, Master Raj made a surprise announcement. Everyone would participate in a "volume shoot," firing one magazine of 17 rounds at a smaller target. This wouldn't count towards our scores. I decided to use my own rifle to see if my struggles yesterday had anything to do with my weapon.

I hit all 17 shots. My struggles yesterday weren't the fault of my rifle. They were mine.

Lee Jan Jan, one of my section mates, said, "Max, you got 15/16 today? After what happened yesterday? No way, man. I bet Warrant Foo told you to suck on purpose yesterday. Yeah, to show that everyone has strengths and weaknesses. He probably said, 'If even Max sucks at something, then it's okay!' I see what you did. I'm calling your bluff, you asshole."

I laughed.

"But really, congratulations," he said, grinning.

BRING ON THE NIGHT

There was another long wait for nightfall.

The night shoot began at sunset. For some reason, I was more nervous than I was in the morning. I had returned Aneirin's rifle. I was using my own weapon, the one which I'd struggled so badly with yesterday. I wondered if today's success was a fluke. *Maybe I just got lucky. Maybe I really am a terrible shooter.*

Master Raj said, "Firers, watch your front."

I looked away, breathed in, and breathed out. I focused on my fear, concentrating until the fire in my chest subsided to a tiny flame, a match's flicker in a windy night.

I hit all 16 rounds. Overall, I shot 31/32.

It's absurdly relieving to know that I'm not helpless with a weapon. I was tremendously grateful to have been able to re-shoot today. If, for example, I had shot well enough to barely pass yesterday, my struggles would have bothered me for a long time. Now, at least I know I can do it.

Though I had 31 hits, because I'm a re-shooter, my official score will be the minimum passing mark of 16/32. I really don't care.

Our 5-tonners rumbled back into NDU at 2230. We returned our arms. Our water parade was at 2315, and then lights-out. We have a great deal for tomorrow: after breakfast, we can sleep until lunch and then rest all afternoon. Our 16-km route march begins at 1800.

THURSDAY, MAY 16

We woke up for breakfast and then slept till noon. It was amazing. We drew our weapons in the afternoon for rifle cleaning.

Our section sat together during the rifle cleaning. At one point, Jerome was pretending to use his magazine as a phone. I decided to join in. I put my mag to my ear and said, "Jerome, are you there?"

Everyone burst into laughter. I didn't think it was that funny, but when I turned my head, I saw why. Instructor Khong was right behind me.

"Max! You think your magazine is a phone, is it?"

"Uh, no instructor!"

"Can you call Warrant Foo for me?"

"No, instructor!"

"Idiots," he muttered, hiding a smile, and moved off.

We slept for another hour before waking up at 1600 for dinner. I was rested and actually excited for the march. Because we were skipping our 24 km, this was our last one. There were whispers, however, that Warrant Jaya had tricked us. That he'd fooled us into thinking this was 16 km when it would actually be 24.

I asked Brendan, "Dude, do you think we'll have to do 24 km today?" His eyes widened. He said, "I'm not sure, man. It seems weird that we only have to do 16. I hope not, though."

The march began at 1800. Instructor Khong led the way.

THE FIRST FOUR KILOMETERS

We completed the first four kilometers in the usual 45 minutes. I talked to Brendan along the way. To keep us preoccupied, we were trying to think of awful safety slogans. Around NDU, there are several banners with safety slogans, and boy, are they awful. Some examples:

"Watch out for the heat or you will be hit."

"If you drink beer, don't put on your dive gear."

"Look ahead or lose a head."

"If we make safety a priority, we will not have any casualty."

I mean, just ridiculous stuff. The last one doesn't even rhyme. Brendan and I tried to come up with equally terrible phrases.

"In order to be safe, wear tights, so you don't chafe," I offered.

"Be careful on your dives so you don't lose your lives," Brendan said.

I countered, "Always check for extra rounds, so you won't end up underground."

Jeffrey Hoe overheard us. He contributed, "Don't be lazy, always remember safety."

"Hydrate, or dehydrate," Kok Yew chipped in.

I said, "You can't spell safety without S-A-F." I thought for a moment, then said, "Hey, shit, that's actually pretty good. I like that one."

Doing stupid stuff makes the march much easier. The key is to not think.

THE SECOND FOUR KILOMETERS

I walked with Elston, Sean and Brendan. It was unpleasant but bearable. The progressive training has been effective. Each march has been easier than the last.

HALFTIME

Eight kilometers down, eight to go. We sat down for our night snack. Cake lemon and cookies sesame have never been so delicious.

We also received "oral rehydration packets." They were gross. They're supposed to be orange-flavored, but they taste like salty sand.

Warrant Jaya arrived. He spotted Aneirin and barked, "Flynn! Stand up. What does loyalty to country mean to you?"

Aneirin swallowed noticeably. "Um, uh," he began, before rambling semi-incoherently about choosing to side with Singapore if Singapore and England ever "get in a fight."

Warrant Jaya cut him off. "Country before self. That's what it means. Right? No need for such bombastic shit. Country before self, that's all!"

"Uh, right, sir," Aneirin said.

Master Dennis shook his head. He exclaimed, "Seven As, but no common sense!"

I was trying not to burst into laughter. Helmethead was sitting next to me, though, and he was cracking up. He's got a silly high-pitched laugh—he sounds like a little girl. As Helmethead giggled to himself, Warrant Jaya looked over and snapped, "Can laugh properly or not?"

After that, I couldn't hold it in.

After the break ended, I mustered next to Aneirin and Ivan. As we stood in file, Master Dennis saw the three of us. He said, "Oi, all the strong ones are together. Split up!" He moved me to the very front, directly behind Instructor Khong, and shifted Ivan to the back.

THE THIRD FOUR KILOMETERS

I was right behind Instructor Khong. Marching is much harder when you aren't with friends.

THE FOURTH FOUR KILOMETERS

I wanted to sneak back to the middle of the pack, but Master Dennis was nearby, so I stayed near the front.

We continued our long, slow plod through Sembawang Camp. As we circled back to NDU for what felt like the hundredth time, I considered dropping out. I really did.

My right ankle was a little sore. I thought, *I could exaggerate my ankle pain. Limp for a while, then shake my head and drop out to the side.* It was an alluring thought. Then, I had another: *Is that who you want to be?* I stifled the temptation.

Finally, we were headed back to Block 29. We snaked through silent roads. Murmured chatter swelled amongst us. We knew we were finishing.

FINISHED

We stood in file before Block 29, aching and sleepy. Instructor Khong looked stern. We were waiting for the most heavenly of commands: "Field packs down."

A minute passed, then another. Tiredness dampened my confusion, but I was still unsure why we remained standing.

Instructor Khong shouted, "39! You have eight more kilometers to finish your 24. Do you want to do it?"

Oh, God. No way.

There was only one answer we could give. "Yes, instructor!" we called.

"Do you want to do it?"

"Yes, instructor!"

"Right face! Follow me." Instructor Khong turned and strode back up the asphalt, leading away from 29.

I exhaled. Our premonition had materialized. I tried to shake my resolve awake. *Alright, eight more kilometers. Then you can sleep. Here we go.*

We marched one hundred meters, turned around, and returned to Block 29.

FINISHED.

We returned our rifles and secured.

Some didn't shower. Some didn't even change, just plopping down on their beds. I stretched briefly and then rinsed off. I didn't bother with soap.

Three guys didn't complete the march due to supposed illness. None were noticeably ill. I lost respect for them.

It's 0140 right now, our latest lights-out ever.

FRIDAY, MAY 17

In the morning, we started rehearsing for next week's Passing Out Parade (POP). It's a short ceremony commemorating the end of Basic Military Training. The rehearsals consist mainly of marching and rifle drills.

The majority of National Servicemen complete their BMT on Pulau Tekong. After passing out, they are assigned to various postings scattered across the SAF. If we make it through CDC and graduate as divers, however, we'll stay in NDU for our entire National Service. For this, I'm glad. It would be great to stick with the same batch for two years.

RECRUITS

We rehearsed for our POP throughout the morning and afternoon. We booked out after dinner. Before we left, Warrant Foo spoke to us, giving us the obligatory weekend warning to not do anything silly while in uniform. "Your book-in timing is 2200," he concluded.

Rajeev, our IC, stood up. He said, "Okay, hear that? Everyone, meet outside at 2100."

Warrant Foo said, "No, I said 2200, not 2100. IC, open your ears!"

Rajeev explained, "No, sir, we're going to meet at 2100 to muster outside first."

"What? Why do you meet outside first?" Warrant Foo was incredulous. He had no idea that before booking in, we always gather outside Sembawang Camp an hour early and walk in together.

"But sir, you told us to do that before our first book-in."

"Bloody hell. That was only supposed to be for the first week! You've been booking in together this whole time?"

We exploded into laughs and groans. We couldn't believe it. *We've been booking in an hour early every week? Are you kidding me?*

Warrant Foo said, "You meet one hour earlier before booking in, every time?"

"Yes, sir."

Warrant Foo shook his head. "Recruits."

WEEK
NINE

SUNDAY, MAY 19

I reached Sembawang Camp at 2140, giving me 20 minutes to walk in. Water parade was at 2200 under Instructor Russell, who didn't even make us drink a full bottle.

More importantly, however, today was Jay Yang's birthday. Once we were dismissed for the night, Aneirin and Elston talked to Jay, stalling him. I dashed upstairs and filled a bucket with cold water.

I crept to the balcony and looked down. Jay was directly below. I signaled to Elston. He and Aneirin said goodbye and walked away, leaving James all alone.

"Happy birthday, asshole!" I shouted, emptying the bucket.

Somehow, without looking up, Jay dodged the waterfall. He took two quick steps and avoided my watery onslaught. It was incredible. I still don't know how he did it.

"What the heck?" Jay said.

"Max, you missed!" Elston cried.

Half the batch was still downstairs. They made up for it by swarming Jay and pouring their water bottles over his head.

Don't worry, though. We *ang mohs* always get our

revenge. After everyone was done wishing Jay a happy birthday, he went upstairs. As Jay turned a corner, Aneirin launched a full bucket of water into his face.

AWARDS

I can't believe this is our last week of BMT. I already feel I've learned so much. One thing we haven't learned, though, is who our award winners are. Two awards will be presented during our graduation from BMT: Best Trainee and Best PT.

Best PT goes to the Best in Physical Training, the fittest recruit. I'm pretty sure this will be Ivan. He's easily the fastest runner, and pretty strong as well.

Best Trainee goes to the overall best performer.

I'm conflicted. I'm naturally competitive, and when I'm being honest with myself, a little insecure. I want to win, because I want to be the best, but also because I want to validate myself. I want to prove to myself that I'm worthy. But I'm also unsure if I deserve it.

MONDAY, MAY 20

Master Dennis was waiting downstairs at Block 29. We were mustered in PT attire, ready for our regular Monday morning endurance run. I just wanted to get it over with.

Master Dennis eyed us. "Man, I don't want to run today," he said.

Murmurs of hope rippled amongst us.

He thought for a couple of seconds, then asked, "Do you all want to play games today?"

"Yes, sir!" we roared.

"Okay," he said, with a shrug. "Let me call Warrant Foo to see if it's okay."

We churned with anticipation. I forced myself to remain unexcited. I was skeptical; this could easily be a trick. If I don't expect anything good, I won't be disheartened if it's taken away. I need to be ready.

Master Dennis pocketed his phone. He said, "All right, 39, games day is on. Everyone, basketball courts, let's go!"

We cheered.

Our options were basketball and soccer. I picked basketball. We played 10-minute full-court contests. My teammates were Aneirin, Raymond, Brendan, and Jia Ming. We won both games we played. I scored and rebounded quite a bit. It was fun. Aneirin said he played basketball in school, but he didn't score at all. Once, he shot the ball over the backboard.

WHEN THE RECRUITS GO MARCHING IN

Back to reality: more rehearsals for Thursday's POP. After our Passing Out Parade, we have Friday off.

We jogged back to our cabins, changed into vest slacks, and drew our rifles. We were feeling good from the morning's games. As we returned to the football field, we were talking loudly, marching out of step, and not in file. Tragically, CO Dive School happened to be walking by. CO scolded us and ordered us to march back to Block 29, turn around, and march back properly. He also informed our instructors.

Our second trip to the football field was in complete silence. When we arrived, we were whacked. Master Raj even yelled at us for the first time. We hit the fence three times with our rifles overhead. I felt lucky we only had to do it three times.

THE REVEAL

Before we resumed rehearsals, Master Raj said, "Your behavior this morning was unacceptable. This is the fault of the ICs. You will now be changing ICs. We need a new volunteer for company IC."

None of us raised our hands.

"Okay, Ivan, come here. Thanks for volunteering.

Does anyone know why he's IC?"

Somebody yelled, "Best trainee!"

Master Raj looked genuinely surprised. He said, "Yes, how did you know? Ivan will also be receiving Best PT. Two awards."

The class broke into applause. Ivan stood up, smiling hugely.

So there it was — the big reveal. Ivan was definitely the favorite for Best PT, but I didn't expect him to snag *both* awards. He's inarguably one of our top performers.

I thought Aneirin deserved Best Trainee, though. Aneirin said he felt okay, but I think he's a little disappointed. After the announcement, Master Dennis shook Aneirin's hand and told him that he's done extremely well. He has.

I'm not sure how I feel about the awards. It's strange to admit, but I'm actually relieved that I didn't win. I don't think I deserved either award. Winning would have been a tremendous thrill, but a thrill streaked with guilt. Ivan and Aneirin are both more deserving than I am.

I'm going to keep working, though. I'm going to get better.

I'm setting my sights on the Combat Diver Course.

⋯⊶ ⊷⋯

Following lunch was further practice for our POP. The ceremony is only a half-hour long, but as it's a reflection of how well we've been taught, our officers want it to be perfect.

In recognition of our completion of BMT, the instructors gave us a special deal. They granted us "liberty"—a few hours outside camp. We can leave camp and spend the evening outside before returning for lights-out. We have to wear smart four, though, so we must behave.

Aneirin, Alistair, Nathan, Jay, Zhi Heng, Siu Hoi, Keshav and I headed to North Point, a nearby mall. We got bubble tea, roamed around, and had dinner. We headed to a *pasar malam* and ate some more, returning to NDU at 2230. It was a great night.

After water parade, Aaron told me he thought I could've won either of the two awards. I thanked him.

I did push-ups and dips before going to bed.

I was in the middle of a set of push-ups to failure when Benjamin came over. He encouraged me to push farther. When I finally flopped, he asked, "How many did you do?"

"Fifty-nine," I said. He giggled.

My push-ups are still second-tier. I'm nowhere near as good as the strongest guys in the batch. Shawn Ong, Thomas, and a few others can easily crank out a hundred. Benjamin can do over 80. I need to get better.

Thomas is deceptively fit. He's skinny and doesn't have a lot of muscle definition, but is exceptionally strong. He told me during our last route march that his shoulders weren't sore at all. Meanwhile, mine were on fire.

OPTIMISM

Aneirin said that for the last couple of days, he's felt a sharp pain in his chest when he inhales deeply.

I said, "Maybe you have a broken rib and it's going to puncture your lung and you'll die."

He replied, "Maybe."

CLOSING TIME

I'll be sad when BMT ends. A few guys won't "class up," or move on to the Combat Diver Course, dropping out voluntarily or due to injury, lack of fitness, or lack of water confidence. We'll also inherit a new set of instructors, which means saying goodbye to Warrant Jaya, Warrant Foo, Master Raj, Master Dennis, and the sergeants.

Our Platoon Commander for CDC will be someone named Warrant Hans. Some of the guys in Batch 38 told us he's fierce but fair. Warrant Foo said to us once, "Hans is a good man. He's fair. I don't say that about many people, but Warrant Hans is a good man."

We look forward to meeting him.

BACK IN THE DAY

Aneirin and I were talking about our childhoods. He was born in Singapore but grew up in England. He only moved here when he was 16.

He said, "Growing up, I never knew I'd have to do NS, you know. I didn't think I'd have to do it."

"What? But you were born here, right?"

"Yeah, I was, but we left for England right after. My parents didn't know I had to do it, either. When I got that enlistment notice, it turned my whole life upside down. I grew up in England, I had a girlfriend, I had friends, and suddenly, boom — you're going to Singapore."

He enrolled in Victoria Junior College. Growing up in England, he drank, smoked, and got into fights. He even ran away from the cops a couple of times. I laughed. I was totally surprised to hear this. "Sounds like you were a thug," I said.

"Yeah," he laughed, "you probably wouldn't have wanted to be my friend back then."

I, on the other hand, was an undersized, baby-faced, innocent youngster. (Okay, I still have a baby face.) When I was 15 years old, I was 160 cm tall and weighed 50 kg. I only started drinking when I was 18, and even then, my curfew was midnight. I've never smoked anything in my entire life. It just doesn't appeal to me.

Now, though, Aneirin and I are quite similar. We'll have a few drinks on the weekend, but there's no fighting or police evasion.

TUESDAY, MAY 21

One of the requirements for classing up to CDC is achieving a minimum of IPPT silver. Today there was an IPPT for those who still hadn't made it. Those with silver or gold were allowed to rest. Thankfully, everyone passed.

We ran through more POP rehearsals until lunch. They worked us hard and long. When the instructors mercifully released us at 1300, I was starving. We all were. As we headed off to the cookhouse, Ivan called, "Guys, we need two volunteers to stay back and guard the rifles." Naturally, nobody volunteered. Ivan said, "Max, Flynn, you two stay back."

With a straight face, I said, "Yes! Okay, good, good. Thanks, Ivan! I wanted to do this. Awesome choice. Really, top-notch." Ivan smiled and winked. Aneirin yelled, "Pick on the *ang mohs* some more!"

Fifteen minutes later, two guys came to switch with us. Finally, we could eat.

Aneirin and I started for the cookhouse, but were intercepted by Instructor Khong. He had secret news.

He said, "Max, you know something? You were supposed to get Best PT."

I stopped walking, jaw hanging open. "What?" I managed.

"You actually scored better than Ivan. The problem was, you messed up one evolution. Think. Which evolution did you really struggle on?"

My brain motored. *What could it be? Swimming, running, IPPT…oh, no.*

"Oh, shit!" I exclaimed.

"Yes," Instructor Khong said, grinning. "I don't even need to say it, do I?"

"Wait, which one was it?" Aneirin asked.

I pretended to fire a rifle.

Instructor Khong said, "Yes. That's right. If you hadn't screwed up your shooting, you would've gotten Best PT."

"Oh, man!" I exclaimed.

Then I laughed. What else could I do?

"Thank you for telling me, Instructor," I said. He nodded. Aneirin and I went to eat.

⌖ ——— ⌖

We were supposed to continue rehearsing after lunch, but it rained for three hours. We were trapped under the grandstand. The instructors refused to let us sleep. After an hour, no matter how hard I tried, I couldn't stay awake. I felt myself nodding off when I heard Master Dennis shout, "Oi, *ang moh*! Wake up before I smack you!"

I snapped up. "Yes, sir!"

Approximately 12 seconds later, I was asleep. When

Master Dennis saw me, I had to hold my rifle overhead for five minutes.

Around 1700, the rain deigned to slow, and we squeezed in one more rehearsal.

At one point in the ceremony, the national anthem will play, and we sing along loudly. Just before we started, Master Dennis yelled to me, "Oi! You better know the words to my country's anthem!"

Luckily, I'd memorized the lyrics before enlisting. Plus, Master Dennis, it's my country too.

Aneirin, on the other hand, went to school at VJC [Victoria Junior College], where they play the national anthem every morning, and still has no idea what the words are.

INTRUDER ALERT

Our rehearsals finished. Warrant Jaya gathered us and made a serious announcement. He said, "I have some news. NDU received a letter from MINDEF today. There have been intruders spotted in the Sembawang area. Armed intruders. All of the SAF is on high alert."

I was frozen. *No way!* This was scary shit!

"They don't know who the intruders are, but they know they're out there, and they're dangerous. NDU has been tasked with a recon mission. NDU's job is to patrol the area, to search for these intruders."

We cheered, proud that NDU was given an important assignment.

"Actually," Warrant Jaya went on, "this mission is for you."

We collapsed into silence.

"The mission calls for you to cover an area on foot. You will walk around the Sembawang area, searching for these intruders, for a total distance of 24 km. Do you accept this mission?"

We burst into laughter. Shit — this was a route march! So we *had* been tricked. The 16 km wasn't really our last one. We do have to do our 24-km march. The instructors had fooled us.

We started to moan and groan, but Wei Qiang shrugged and said, "At least got storyline. Not bad eh."

The march will be tomorrow evening, beginning around 2100 and ending around 0500. We'll march through the night.

WEDNESDAY, MAY 22 – THURSDAY, MAY 23

We rehearsed for our POP from morning till lunch, then were allowed to sleep until 1600, resting up for the big march. We awoke for dinner at 1630, then rested and prepared our field packs until 1900.

Oscar borrowed my boot-polishing brush last week. Today, I found out he lost it. I was ready to be mad at him, but then he offered me gummy worms.

SUIT UP

We received new number fours to wear during our Passing Out Parade. Instead of the traditional green Army number fours, we'll wear Navy number fours, built of blue and white. They look awesome.

We're setting off for the march in an hour. Here we go.

THE FINAL ROUTE MARCH

It sucked, and that's about all I have to say about it.

PLEASE, MAX, TELL US HOW IT WENT!

Okay, fine. It was a slog. We did the first 6 km inside Sembawang Camp, then completed the remainder of the march in public.

Our first break came in Sembawang Park. It was about 0130. Aneirin and I sat down. Night snacks were distributed. I turned the packet over to find that our snacks were fantastically named "Cookies Chinese." They had an odd, spicy taste.

"Were these made *by* Chinese people…or *of* Chinese people?" I wondered.

"Shut up," Aneirin said.

We carried on, plodding thickly through the night. We carried on with our usual marching pattern of completing 4 km in 45–60 minutes, then having a short rest.

By 0230, at our next break, I was exhausted. Those of us who needed a bathroom break were allowed to piss over a railing into a rain sewer. We had some fun with it. As we walked over to the railing, George shouted, "Everyone, form up your firing line!"

Jonas Koh yelled, "Remember your muzzle discipline! Don't point at your buddy."

As we laughed and peed, George yelled, "IA, IA, IA!"

By 0330, I was sleepwalking. My eyes would close for a few seconds, I'd stumble or shake myself awake, carry on for a few steps, and then my eyes would close again. This actually made the march easier—when I was asleep, I didn't feel tired. I think I even dreamed a few times.

We continued around Sembawang and Yishun, trudging through the small hours. It was true sludge. It was definitely easier to march in public than in camp.

The changing scenery helped. We weren't allowed to sing, however, which was a slight handicap.

Sidewalks and streetlights blurred into continuous streams of gray concrete and dull orange. At some point, I jerked awake and realized we were covering old ground. A surge of energy streamed through my veins. We were going back.

I've never been as happy to book in as I was when we marched back through Sembawang's sweet, sweet gates. It was 0530. We stomped into NDU, dropped our field packs, and were officially done with our 24-km route march.

Most of us wanted to sleep immediately, but the instructors ordered us to have breakfast. We ate quickly and returned to Block 29.

I hit my bed just as slivers of sunlight were edging over the horizon.

THE LAST DAY

We were allowed to sleep until 1000. When we woke up, I realized that this was it. Tonight was our Passing Out Parade. Our last day of BMT had arrived.

None of us wanted to rehearse again, but that's what we did, doing a full run-through before lunch and then a final rehearsal in the afternoon. We changed into our Navy smart fours and jockey cap, standing tall in gray and blue, rifles at the ready. We looked good. Really good.

During dinner, gravelly thrums of thunder rolled over us. *Oh, no.* If there's rain, our parade will be moved from the football field to the multi-purpose hall. If this happens, there will be few drills and no marching — we'll just stand there while the instructors conduct the ceremony. In short, all of our rehearsing will have been wasted.

Lightning crackled across the darkening sky.

We waited in an elevated training shed by the pool. We were hidden from the grandstand so arriving guests couldn't see us. I tried to stay hopeful. The ceremony was scheduled to start at 1900. At 1830, the clouds held steady, but at 1845, we received the word: our parade was on.

We mustered on a road around the corner from the grandstand. We were split into two contingents, Platoon 1 and Platoon 2. Each platoon was arranged by height, with the tallest on the ends and the shortest in the middle.

Everyone was nervous.

From the field, we heard Master Raj scream the command to march: "*Dari kiri, cepat ja-lan!*"

Ceremonial music flowed through the speakers and we marched in, wearing our Navy number fours and combat boots, uniforms crisp, rifles at the ready. As we stepped onto the field, I glanced to the right and was shocked. The grandstand was crammed. The crowd was twice the size I'd expected.

We reached our positions and proceeded to *hentak kaki*, marching on the spot. As the last drum beats sounded, we turned to face the grandstand and stopped in synchrony as the song ended. The audience applauded.

"*Baris, julang senja-ta!*"

We snapped our rifles to our right sides, upright, and whipped our left arms down. Master Raj handed over command of the parade to Warrant Foo.

Then came the presenting of the awards. 3SG Wayne read over the loudspeakers, "There will be two awards presented tonight, the Best PT award and the Best Trainee award. First is the Best PT award."

My mom and 16-year-old brother, Bart, were in the audience. I told them where to sit so they could spot me in the file. They were directly in front of me. I had also told them about the two awards that would be handed out. I'd forgotten, however, to tell them that I wouldn't be receiving either of them. *Shit,* I thought. *They're hoping I'm going to win.*

"The Best PT award goes to…Recruit Ivan Tan."

Amidst the cascading applause, Ivan jogged up to the podium to receive his plaque. He placed it down, and returned to his spot in the file.

"Next is the Best Trainee award. The best trainee is an all-rounder. He is a great leader as well as a cherished batch boy. The Best Trainee award goes to Recruit Tan Qing Yu, Ivan."

There were scattered gasps, and then more applause.

The crowd was surprised to see the same recruit win both awards. I saw my brother grit his teeth and look away, disappointed that I'd lost. I was touched. Ivan ran to the podium and repeated the procedure.

CO Dive School gave a speech on the importance of National Service and our service to the nation. When that ended, Instructor Wayne announced, "Next, you will be witnessing the march past."

We turned to the right, facing the direction from which we'd marched in. As the ceremonial music rang out, we marched off, turned around at the end of the field, and marched past the grandstand once more. We turned around once again and then passed the grandstand a final time before exiting the field.

Around the corner and out of sight, we put down our rifles, put our berets in our pockets, and scrambled back into formation. Ivan stood alone in between the platoons.

We marched back onto the field, this time, without music. A guy named Richard Lum called out the timing. "Left, left right. Left, left right. Left right left right left right left."

We halted on Ivan's command. The audience rose for the national anthem, which we sang with verve if not skill. We then recited the SAF pledge, repeating each line after LTA Gabriel, our right hands over our hearts.

We, members of the Singapore Armed Forces,
Do solemnly and sincerely pledge that,
We will always bear true faith and allegiance,
To the President and the Republic of Singapore.
We will always support and defend the Constitution.
We will preserve and protect
The honor and independence of our country
With our lives.

"With our lives, SIR!" we roared.

Instructor Wayne then invited our parents to come down to the field, remove our jockey caps and place our berets on our heads, symbolizing the end of Basic Military Training. Instructor Wayne said, "I'm sure your son has been picturing this moment for a long time, so please head down to the field and reunite with your loved ones."

I was moved when my mom came down, teary-eyed. I handed her my beret and she placed it on my head. I took photos with her and with Bart.

When all the parents had returned to their seats, only one portion of the POP remained. His voice steadily rising, Instructor Wayne announced, "Ladies and gentlemen, our newly trained soldiers!"

We threw our jockey caps into the air and yelled, "39, HOOYAH!" Celebratory tunes blasted through the night and we leapt around in glee.

Nine weeks of Basic Military Training was over. We'd

done it. We are recruits no longer — now, we're privates.

Platoon 2 ran over and joined us. We hoisted Ivan upon our shoulders, chanting his name. After dancing and jumping around for around five minutes, just as the furor was fading, someone yelled, "Hit the fence!" We all turned and sprinted for the fence, hollering like madmen.

Eventually, we gathered at the basketball court for photos. I stayed for 20 minutes, getting pictures with as many guys as I could.

Master Dennis approached me. He shook my hand and said, "Good job, Max. You're definitely top three in the class. All the best for CDC. One thing I have for you — try not to look so sleepy all the time. Anyway, good job." I thanked him.

It was dark by the time we left. As we drove out, I pointed out Block 29 to Bart and my mom, including cabin #02-03, my home for the last nine weeks.

THE END OF THE BEGINNING

As I walked out of NDU, I realized that this was the last time I would leave the Naval Diving Unit as a soldier in BMT. When we return on Monday, we'll officially be in Dive School, in the Combat Diver Course. We'll need to be stronger, harder, better. On Monday, we'll be recruits no more, born again as divers-to-be.

As I stood there on the grass during our POP, beneath the bright lights, with my family in the stands and my new brothers standing alongside me, sweating through my number four but standing firm, grasping my rifle in my arms, I was struck by a flash of rare insight.

Basic Military Training was done. Nine weeks: nine of the longest, toughest, strangest, most revealing weeks of my life. I learned far more than I had in any classroom. I was trained and equipped with the basic skills necessary to serve my country and protect the land upon which I now stood, combat boots planted atop grass and dirt. With luck, these were skills I would never be required to deploy, but they were skills I possessed nonetheless. And best and scariest of all, the greatest challenges lie ahead. The Combat Diver Course looms. My feet edge closer to the precipice. I feel myself looking into a dark stormcloud, thick with thunder and rain, promising bold experience and untold pain. I don't know if I'll be able to endure what comes next. I only know one thing: I want to find out.

And, as I stood there, I smiled. I smiled with triumph but also with sadness, because though the path ahead was perilous, I knew that I would miss this, would miss these boys and these days; knew that I would smile every time I looked back, but that every time I did, this time would be further away; that with each second slipping between us, the days of the 39th Batch would fade and gray; but that someday, in 20 years or 20 minutes, I would look

back and know that these were the best days of my life.
I knew I would miss this.
And I was right.

The author (right) and his teammates take advantage of a rare break from training.

A team powers through the paddle run, in which trainees must stay connected by holding each other's paddles while running long distances.

The 39th Batch gets wet and sandy.

The author and his batch boys recite the SAF Pledge during their Passing Out Parade.

The author with his brother, Bart, and his mother, Lai Fan,
after the graduation ceremony.

The original members of the 39th Batch book out for the very first time.

water, so we wake up early to ~~d~~ eat brkfst & do so. _5am_

• ~~Late~~ At 6am we ~~find out~~ the staff ~~p~~ arrives to lead our workout. SSM unavailable; has to attend to PTP batch enlistment.

• ~~Keep~~ Same PT format: Ran 2k, then brk at courts. 20 pushups, 30 crunches. Run ≈1.5k (~— est.), then courts for 20 pushps, 30 crunches. Run ≈1-1.5k, 25 pushups, 30 crunches.

• ~~ten~~ then do one short jog around then run ~~route~~ ~~go~~ line up on football field. ~~—~~ says we gonna pay for the two mistakes we made!

2) En 1st ~~pay~~ round of running someone (hooyah-ed) Master Chief U66 inst. of Mast Chief NDU.

1) 1st row of runners saw M Chief but didn't hooyah him. Mistake 2:

ABOUT THE AUTHOR

Max West was born and raised in Singapore, where he attended the Singapore American School. He performed his National Service in the Naval Diving Unit of the Republic of Singapore Navy. He is currently an undergraduate at Princeton University, where he studies English and creative writing.